REED'S MOTOR ENGINEERING KNOWLEDGE FOR MARINE ENGINEERS

By

THOMAS D. MORTON

EXTRA FIRST CLASS ENGINEERS CERTIFICATE
C.Eng., F.I.Mar.E.

PUBLISHED BY THOMAS REED PUBLICATIONS LIMITED
SUNDERLAND AND LONDON

First Edition 1975
Second Edition 1978

ISBN 0 900335 52 1

PRINTED BY THOMAS REED AND COMPANY LIMITED
SUNDERLAND AND LONDON

PREFACE

The object of this book is to prepare students for the Certificates of Competency of the Department of Trade in the subject of Motor Engineering Knowledge.

The text is intended to cover the ground work required for both examinations. The syllabus and principles involved are virtually the same for both examinations but questions set in the First Class require a more detailed answer.

The book is not to be considered as a close detail reference work but rather as a specific examination guide, in particular **all the sketches are intended as direct application to the examination requirements.**

The best method of study is to read carefully through each chapter, practising sketchwork, and when the principles have been mastered to attempt the few examples at the end of the chapter. Finally, the miscellaneous questions at the end of the book should be worked through. The best preparation for any examination is the work on examples, this is difficult in the subject of Engineering Knowledge as no model answer is available, nor indeed any one text book to cover all the possible questions. As a guide it is suggested that the student finds his information first and then attempts each question in the book in turn, basing his answer on either a good descriptive sketch and writing or a description covering about 1½ pages of A4 paper in ½ hour.

The author wishes to extend his thanks to Mr. J. Duffy, of the Marine and Technical College, South Shields, for his invaluable assistance.

1978

CONTENTS

CHAPTER 1

BASIC PRINCIPLES

DEFINITIONS AND FORMULAE

Isothermal Operation (PV = constant)

An ideal reversible process at constant temperature. Follows Boyle's law, requiring heat addition during expansion and heat extraction during compression. Impractical due to requirement of very slow piston speeds, etc.

Adiabatic Operation (PV^{γ} = constant)

An ideal reversible process with no heat addition or extraction. Work done is equivalent to the change of internal energy. Requires impractically high piston speeds, etc.

Polytropic Operation (PV^{n} = constant)

A more nearly practical process. The value of index n usually lies between unity and gamma.

Volumetric Efficiency

The proportion existing between the mass of suction air actually contained in the cylinder at the start of the compression stroke and that mass of air evacuated from the cylinder swept volume calculated at normal (or standard) conditions. Usually used to describe 4-stroke engines and air compressors and the general value is about 90%.

Scavenge Efficiency

Similar to volumetric efficiency but used to describe 2-stroke engines where some gas may be included with the air at the start of compression. Both efficiency values are reduced by high revolutions,high ambient air temperature, etc.

Mechanical Efficiency

A measure of the mechanical perfection of an engine. Numerically expressed as the ratio between the indicated power and the brake power.

Uniflow Scavenge

Exhaust at one end of the cylinder (top) and scavenge air entry at the other end of the cylinder (bottom) so that there is a clear flow traversing the full cylinder length, *e.g.* Doxford (see Fig. 1).

Loop Scavenge

Exhaust and scavenge air entry at one end of cylinder (bottom), *e.g.* Polar. This general classification simplifies and embraces variations of the sketch (Fig. 1) in cases where air and exhaust are at different sides of the cylinder with and without crossed flow loop (cross and transverse scavenge).

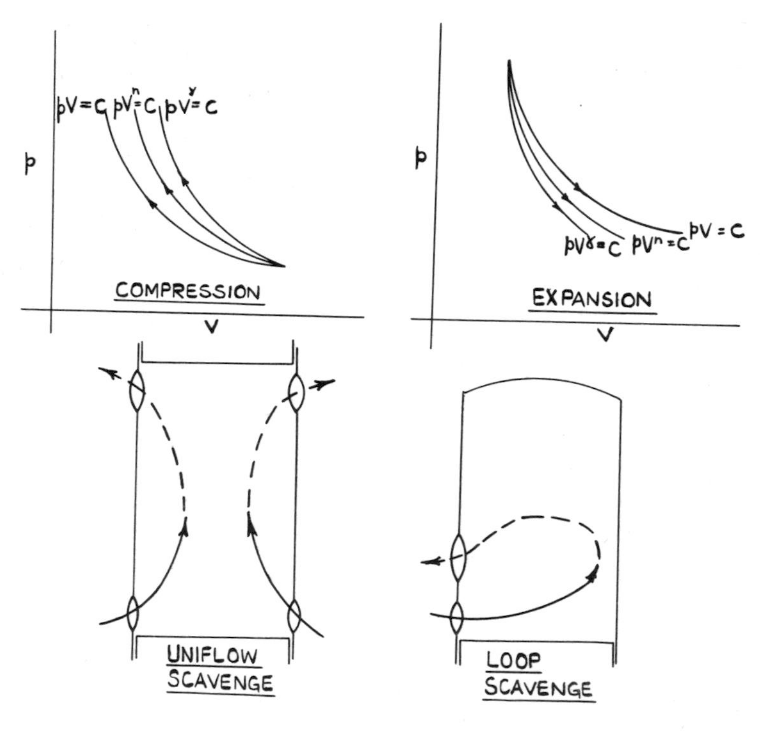

Fig. 1

Brake Thermal Efficiency

The ratio between the energy developed at the brake (output shaft) of the engine and the energy supplied.

Specific Fuel Consumption

Fuel consumption per unit energy at the cylinder or output shaft, kg/kWh (or kg/kWs), 0.21 kg/kWh would be normal on a shaft energy basis for the engine only and 0.235 kg/kWh all purpose (*i.e.* including auxiliaries).

Compression Ratio

Ratio of the volume of air at the start of the compression stroke to the volume of air at the end of this stroke (inner dead centre). Usual value for a compression ignition (CI) oil engine is about 12.5 to 13.5, *i.e.* 8% of the stroke is clearance volume.

Fuel — Air Ratio

Theoretical air is about 14.5 kg/kg fuel but actual air varies from about 29-44 kg/kg fuel. The percentage excess air is about 150 (36.5 kg/kg fuel).

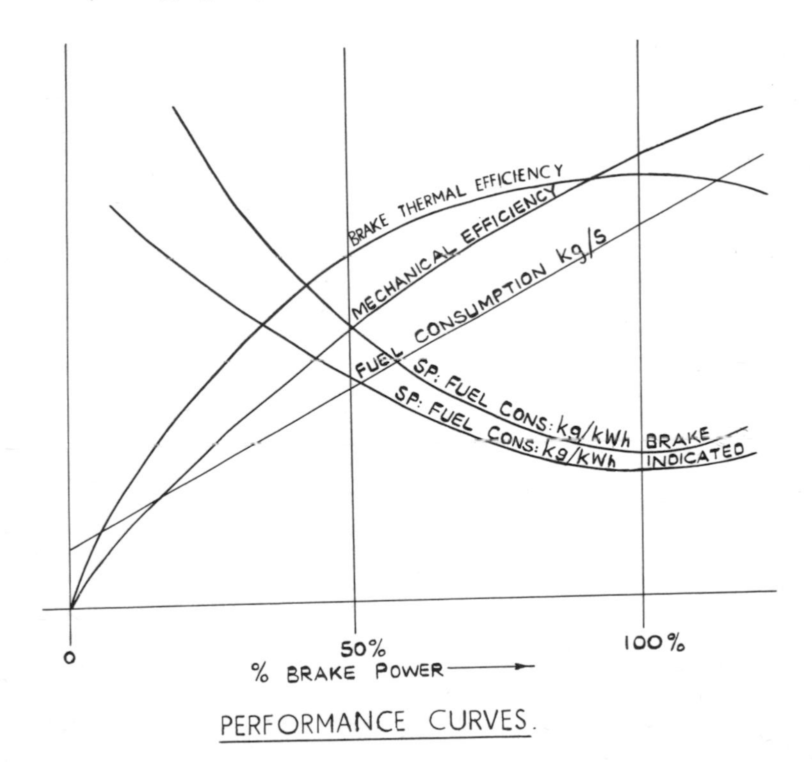

Fig. 2

Performance Curves Fuel Consumption and Efficiency

With main marine engines for merchant ships the optimum designed maximum thermal efficiency (and minimum specific fuel consumption) are arranged for full power conditions. In naval practice minimum specific fuel consumption is at a given percentage of full power for economical speeds but maximum speeds are occasionally required when the specific fuel consumption is much higher. For IC engines driving electrical generators it is often best to arrange peak thermal efficiency at say 70% load maximum as the engine units are probably averaging this load in operation.

The performance curves given in Fig. 2 are useful in establishing principles. The fuel consumption (kg/s) increases steadily with load. Note that halving the load does not half the fuel consumption as certain essentials consume fuel at no load (*e.g.* heat for cooling water warming through, etc.). Willan's law is a similar illustration in steam engine practice.

Mechanical efficiency steadily increases with load as friction losses are almost constant. Thermal efficiency (brake for example) is designed in this case on the sketch for maximum at full load. Specific fuel consumption is therefore a minimum at 100% power. Fuel consumption on a brake basis increases more rapidly than indicated specific fuel consumption as load decreases due to the fairly constant friction loss. In designing engines for different types of duty the specific consumption minima may be at a different load point. As quoted earlier this could be about 70% for engines driving electrical generators.

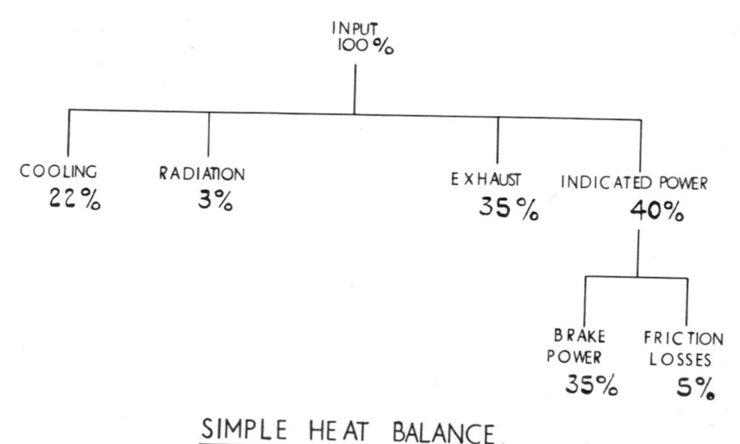

Fig. 3

Heat Balance

A simple heat balance is shown in Fig. 3.

There are some factors not considered in drawing up this balance but as a first analysis this serves to give a useful indication of the heat distribution for the IC engine. The high thermal efficiency and low fuel consumption obtained are still in general superior to any other form of engine in use at present.

A more detailed form of heat balance is shown in Fig. 4 and certain points are now worthy of closer study.

1. The use of a waste heat (exhaust gas) boiler gives a plant efficiency gain as this heat would otherwise be lost up the funnel. This does not of course alter the engine efficiency. The heat extraction efficiency of the waste heat boiler in the given case is about 50%.

2. Exhaust gas driven turbo-blowers contribute to high mechanical efficiency. As the air supply to the engine is not supplied with power directly from the engine, *i.e.* chain driven blowers or direct drive scavenge pumps, then more of the generated power is available for effective brake power.

Consideration of the above shows two basic flaws in the simplification of a heat balance as given in Fig. 3.

(a) The difference between indicated power and brake power is not only the power absorbed in friction. Indicated power is necessarily lost in essential drives for the engine such as camshafts, pumps, blowers, etc. which means a reduced potential for brake power.

(b) Friction results in heat generation which is dissipated in fluid cooling media, *i.e.* oil and water, and hence the cooling analysis in a heat balance should include the frictional heat effect as an assessment.

3. Cooling loss includes an element of heat energy due to generated friction as mentioned above.

4. Propellers do not usually have propulsive efficiencies exceeding 70% which reduces brake power accordingly to the output power.

5. The various efficiencies quoted should be clearly seen from the data on the balance. Specific fuel consumptions as given relate to good modern practice for large direct drive IC engines, with turbo charging.

6. In the previous remarks no account has been taken of the

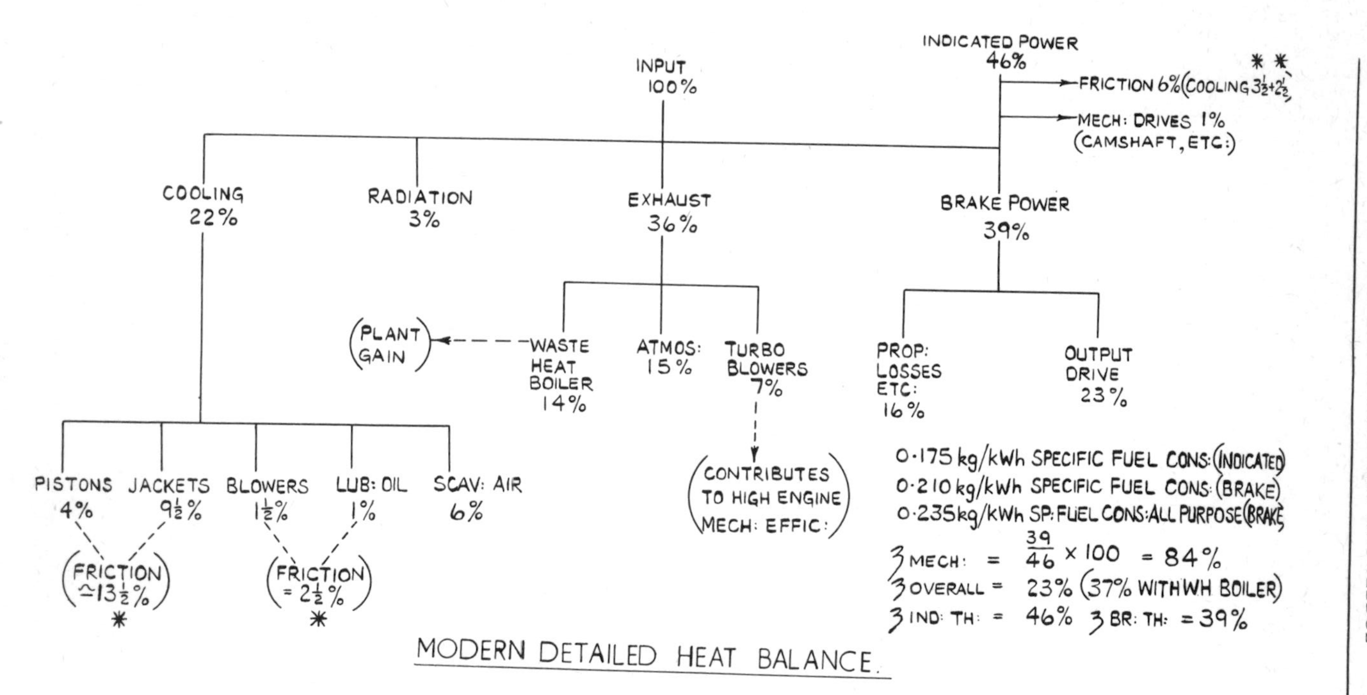
INPUT 100%
INDICATED POWER 46%
FRICTION 6% (COOLING 3½+2½) **
MECH: DRIVES 1% (CAMSHAFT, ETC:)
COOLING 22%
RADIATION 3%
EXHAUST 36%
BRAKE POWER 39%
(PLANT GAIN)
WASTE HEAT BOILER 14%
ATMOS: 15%
TURBO BLOWERS 7%
PROP: LOSSES ETC: 16%
OUTPUT DRIVE 23%
PISTONS 4%
JACKETS 9½%
BLOWERS 1½%
LUB: OIL 1%
SCAV: AIR 6%
(CONTRIBUTES TO HIGH ENGINE MECH: EFFIC:)
(FRICTION ≃13½%) *
(FRICTION = 2½%) *
0·175 kg/kWh SPECIFIC FUEL CONS: (INDICATED)
0·210 kg/kWh SPECIFIC FUEL CONS: (BRAKE)
0·235 kg/kWh SP: FUEL CONS: ALL PURPOSE (BRAKE)
η MECH: = 39/46 × 100 = 84%
η OVERALL = 23% (37% WITH WH BOILER)
η IND: TH: = 46% η BR: TH: = 39%
MODERN DETAILED HEAT BALANCE.

Fig. 4

increasingly common practice of utilising a recovery system for heat normally lost in coolant systems.

IDEAL CYCLES

These cycles form the basis for reference of the actual performance of IC engines. In the cycles considered in detail all curves are frictionless adiabatic, *i.e.* isentropic. The usual assumptions are made such as constant specific heats, mass of charge unaffected by any injected fuel, etc. and hence the expression *'air standard cycle'* may be used. There are two main classifications for reciprocating IC engines, (a) spark ignition (SI) such as petrol and gas engines and, (b) compression ignition (CI) such as diesel and oil engines. Older forms of reference used terms such as light and heavy oil engines but this is not very explicit or satisfactory. Four main air standard cycles are first considered followed by a brief consideration of other such cycles less often considered. The cycles have been sketched using the usual method of P-V diagrams.

Otto (Constant Volume) Cycle

This cycle forms the basis of all SI and high speed CI engines. The four non-flow operations combined into a cycle are shown in Fig. 5.

$$
\begin{aligned}
\text{Air Standard Efficiency} &= \text{Work Done}/\text{Heat Supplied} \\
&= \frac{(\text{Heat Supplied} - \text{Heat Rejected})}{\text{Heat Supplied}} \\
&= 1 - \text{Heat Rejected}/\text{Heat Supplied} \\
&= 1 - mc(T_3 - T_2)/mc(T_4 - T_1) \\
&= 1 - 1/(r - 1)
\end{aligned}
$$

[using $T_2 / T_1 = T_3 / T_4 = r^{\gamma - 1}$ Where r is the compression ratio].

Note

Efficiency of the cycle increases with increase of compression ratio. This is true of the other four cycles.

Diesel (Modified Constant Pressure) Cycle

This cycle is more applicable to older CI engines utilising long periods of constant pressure fuel injection period in conjunction with blast injection. Modern engines do not in fact aim at this cycle which in its pure form envisages very high compression ratios. The term semi-diesel was used for hot bulb engines using a compression ratio between that of the Otto and the Diesel ideal cycles. Early Doxford engines utilised a form of this principle with low compression pressures and 'hot spot' pistons. The Diesel cycle is also sketched in Fig. 5 and it may be noted that heat is received at constant pressure and rejected at constant volume.

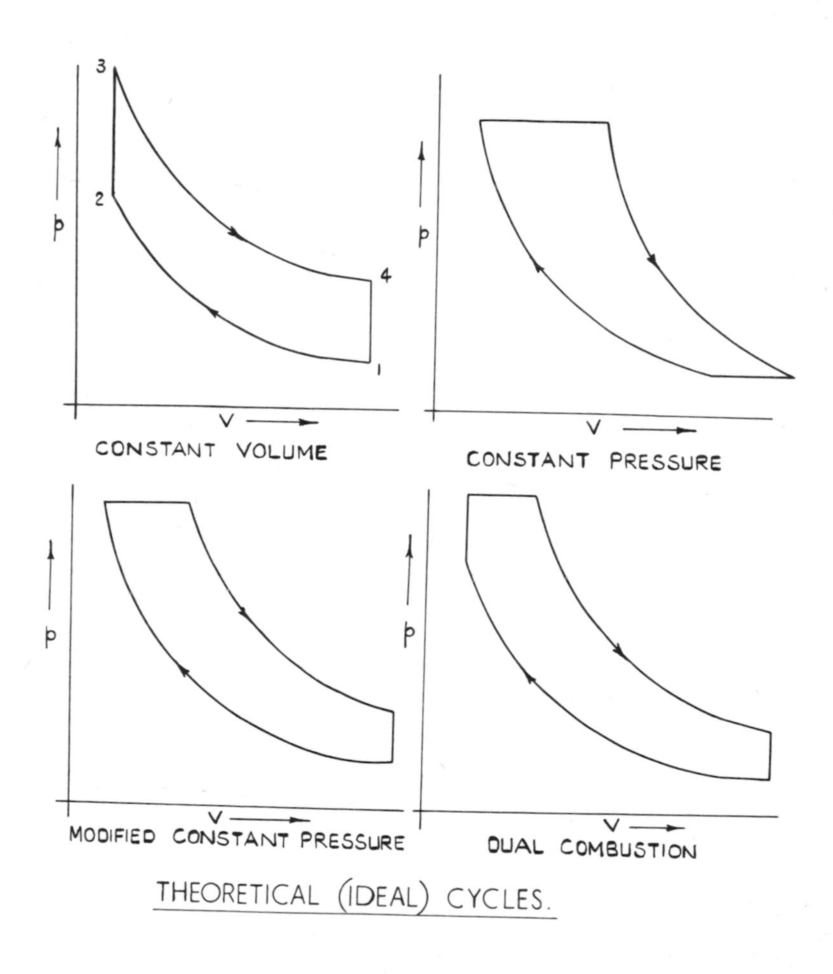

Fig. 5

Dual (Mixed) Cycle

This cycle is applicable to most modern CI reciprocating IC engines. Such engines usually employ solid injection with short fuel injection periods fairly symmetrical about the firing dead centre. The term semi-diesel was often used to describe engines working close to this cycle. In modern turbo-charged marine engines the approach is from this cycle almost to the point of the Otto cycle, *i.e.* the constant pressure period is very short. This produces very heavy firing loads but gives the necessary good combustion. Some typical pressures (and temperatures) around the five cardinal points of such a cycle may give a useful guide and are quoted as follows:
1 bar (15°C), 37 bar (560°C), 47 bar (1030°C), 47 bar (1110°C), 1.5 bar (360°C), r approx: 13:1, Air Standard Efficiency approx: 65%.

Joule (Constant Pressure) Cycle

This is the simple gas turbine flow cycle. Designs at present are mainly of the open cycle type although nuclear systems may well utilise closed cycles. The ideal cycle P-V diagram is shown in Fig. 5 and again as a circuit cycle diagram on Fig. 6 in which inter-coolers, heat exchangers and reheaters have been omitted for simplicity.

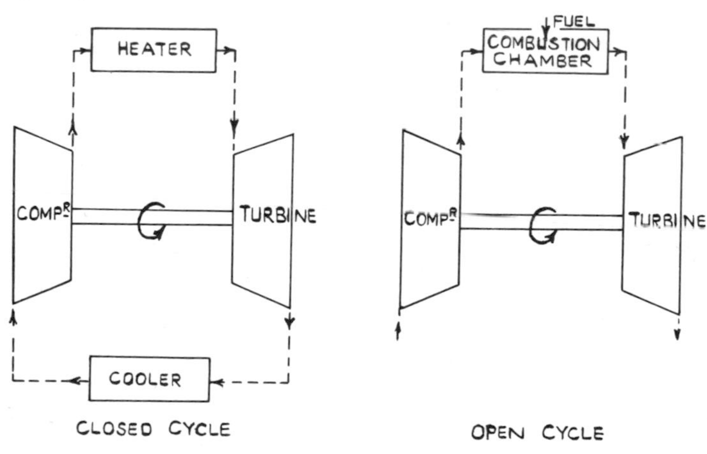

Fig. 6

Other Cycles

The efficiency of a thermodynamic cycle is a maximum when the cycle is made up of reversible operations. The Carnot cycle of isothermals and adiabatics satisfies this condition and this maximum efficiency is given by $(T_1 - T_2)/T_1$ where the Kelvin temperatures are maximum and minimum for the cycle. The cycle is practically not approachable as the mean effective pressure is so small and compression ratio would be excessive. All the four ideal cycles have efficiencies less than the Carnot. The Stirling cycle and the Ericsson cycle have equal efficiency to the Carnot but have little practical value. The Carnot cycle is sketched on both P-V and T-S axes in Fig. 7.

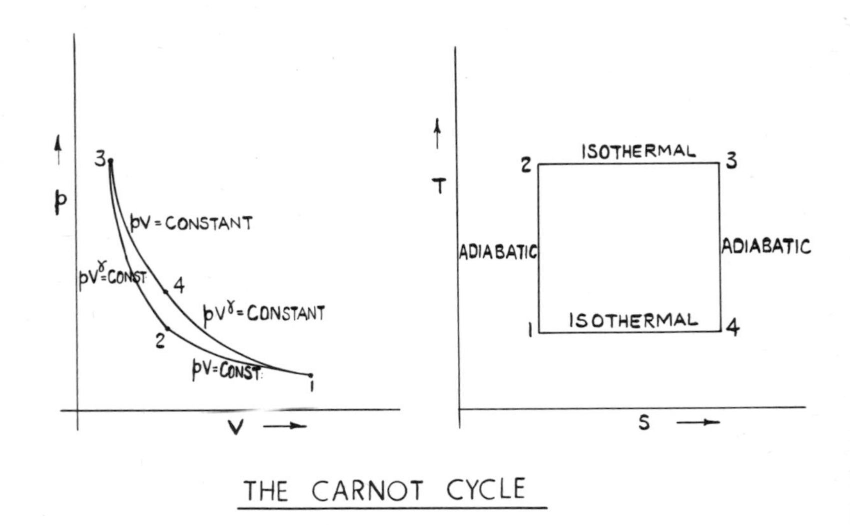

Fig. 7

ACTUAL CYCLES AND INDICATOR DIAGRAMS

There is an analogy between the real IC engine cycle and the equivalent air standard cycle in that the P-V diagrams are similar. The differences between these cycles are now considered and for illustration purposes the sketches given are of the Otto cycle. The principles are however generally the same for most IC engine cycles.

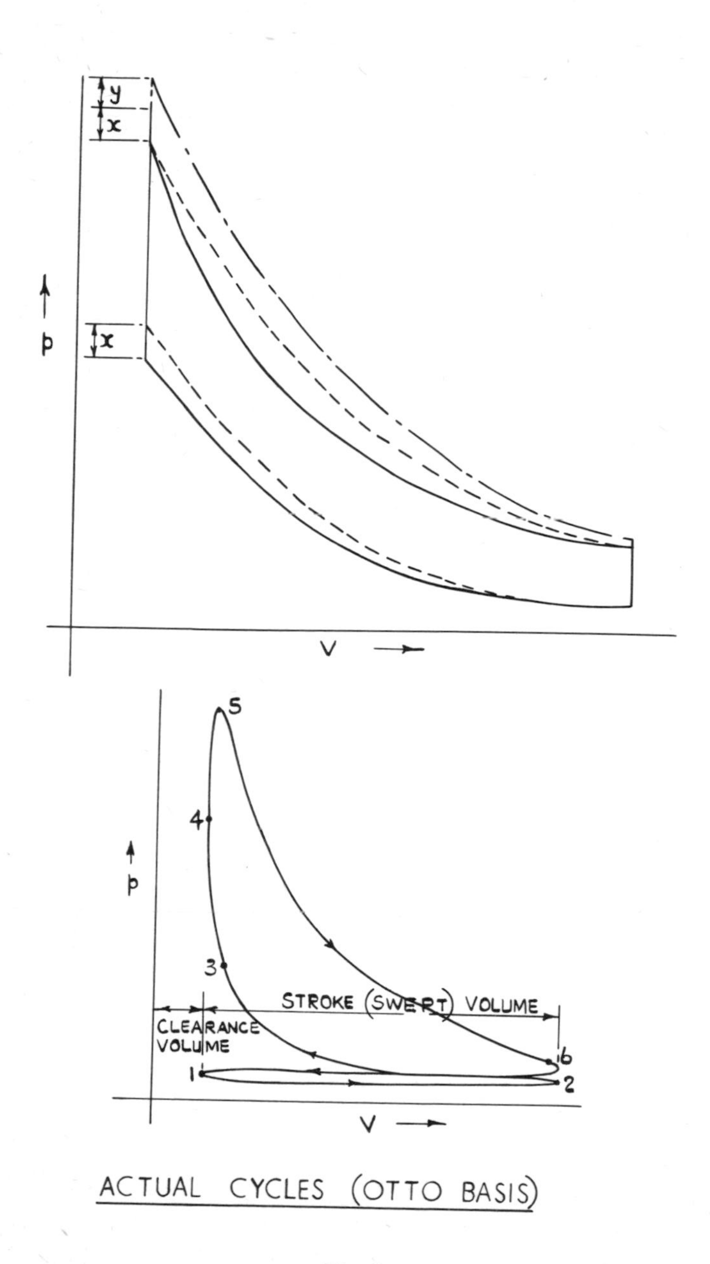
y
x
x
p
V
5
4
3
p
STROKE (SWEPT) VOLUME
CLEARANCE VOLUME
6
1
2
V
ACTUAL CYCLES (OTTO BASIS)

Fig. 8

(a) The actual compression curve (shown full line on Fig. 8) gives a lower terminal pressure and temperature than the ideal adiabatic compression curve (shown dotted). This is caused by heat transfer taking place, variable specific heats, a reduction in γ due to gas-air mixing, etc. Resulting compression is not adiabatic and the difference in vertical height is shown as x.

(b) The actual combustion gives a lower temperature and pressure than the ideal due to dissociation of molecules caused by high temperatures. These twofold effects can be regarded as a loss of peak height of $x + y$ and a lowered expansion line below the ideal adiabatic expansion line. The loss can be regarded as clearly shown between the ideal adiabatic curve from maximum height (shown chain dotted) and the curve with initial point $x + y$ lower (shown dotted).

(c) In fact the expansion is also not adiabatic. There is some heat recovery as molecule re-combination occurs but this is much less than the dissociation combustion heat loss in practical effect. The expansion is also much removed from adiabatic because of heat transfer taking place and variation of specific heats for the hot gas products of combustion. The actual expansion line is shown as a full line on Fig. 8.

In general the assumptions made at the beginning of the section on ideal cycles are worth repeating, *i.e.* isentropic, negligible fuel charge mass, constant specific heats, etc. plus the comments above such as for example on dissociation. Consideration of these factors plus practical details such as rounding of corners due to non-instantaneous valve operation, etc mean that the actual diagram appears as shown in the lower sketch of Fig. 8. This sketch which may be regarded as typical for an SI engine working on the 4-stroke cycle and the following data may be regarded as fairly common practice:

1-2 Induction at 1 bar (temperature at start of compression about 100°C).

2-4 Compression to 12 bar and 420°C with compression ratio about 8:1 and index of compression about 1.33. Spark at 3 causes almost constant volume combustion.

4-6 Expansion to 2 bar and 220°C with index of expansion about 1.28. Peak pressure and temperature at 5 are about 30 bar and 1300°C respectively.

6-1 Exhaust with pressure falling to about 1 bar.

Mean indicated (effective) pressure for such a cycle could be 3 bar and valve timings as follows: induction from 10° before top dead centre for 240° and exhaust from 45° before bottom dead centre for 245°. Timings tend to be early with extended periods due to the high revolutions and indeed are increased further for supercharging.

Typical Indicator Diagrams

The power and draw cards are given on Fig. 9 and should be closely studied. Diagrams given are for compression ignition engines of the 2- and 4-stroke types.

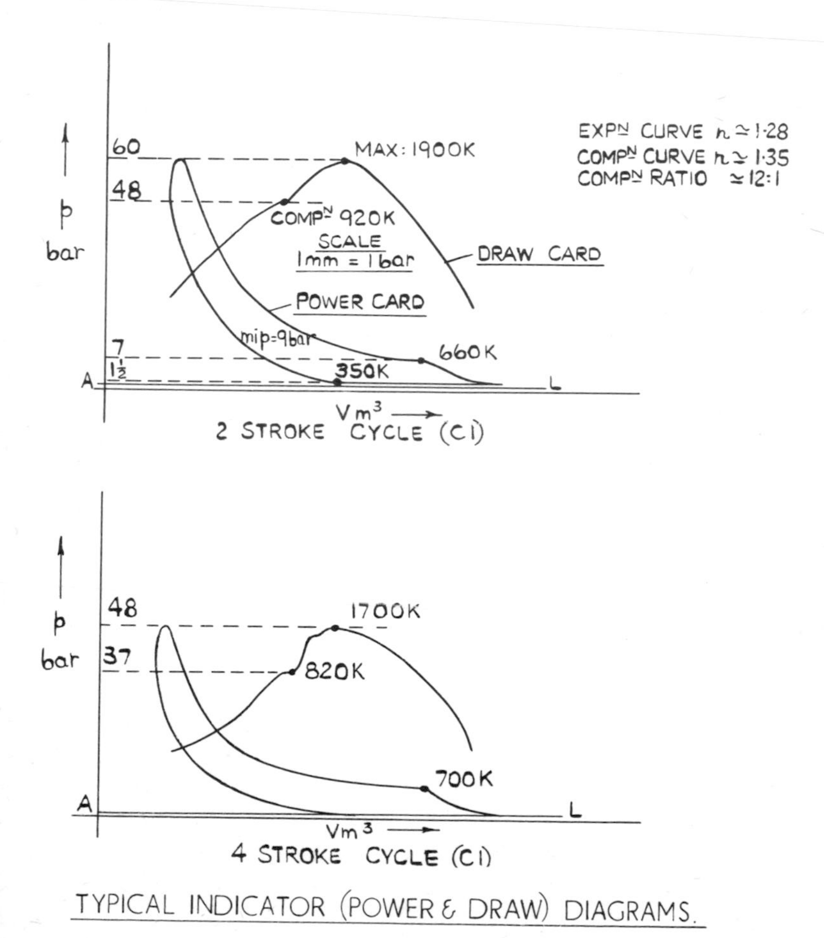

Fig. 9

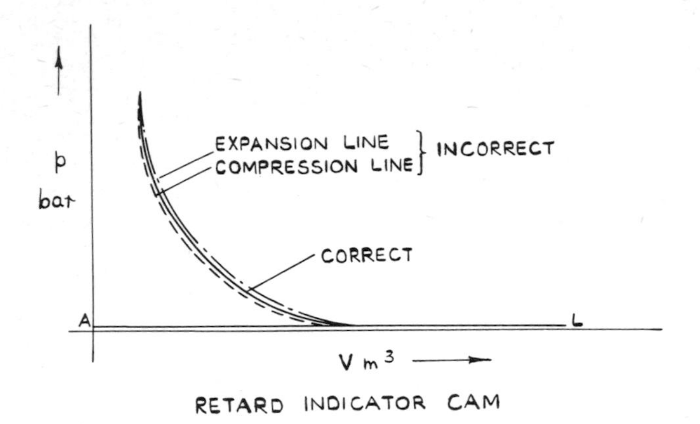

RETARD INDICATOR CAM

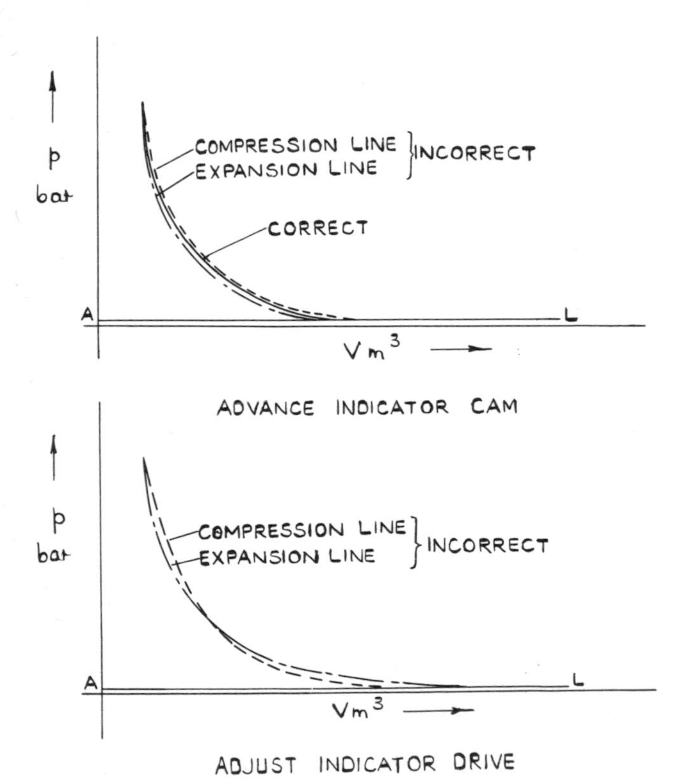

ADJUST INDICATOR DRIVE

COMPRESSION DIAGRAMS (C1)

Fig. 10

Pressures and temperatures are shown on the sketches where appropriate. The draw card is an extended scale picture of the combustion process. In early marine practice the indicator card was drawn by hand — hence the name. In modern practice an 'out of phase' (90°) cam would be provided adjacent to the general indicator cam. Incorrect combustion details show readily on the draw card. There is no real marked difference between the diagrams for 2-stroke or 4-stroke. In general the compression

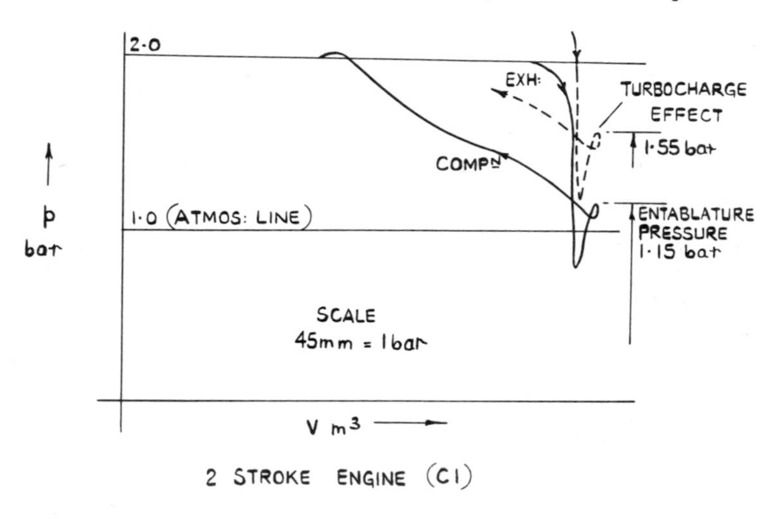

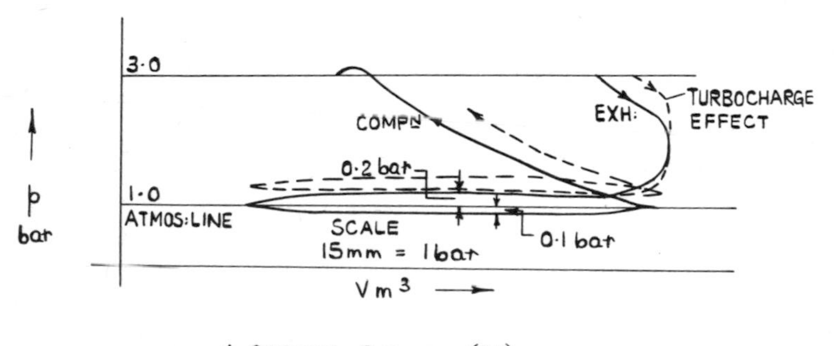

TYPICAL INDICATOR (LIGHT SPRING) DIAGRAMS.

Fig. 11

point on the draw card is more difficult to detect on the 2-stroke as the line is fairly continuous. Exhaust temperature is higher on a 4-stroke as there is no cooling scavenge effect. There is no induction — exhaust loop for the 4-stroke as the spring used in the indicator is too strong to discriminate on a pressure difference of say $\frac{1}{3}$ bar only.

Compression diagrams are given also in Fig. 10; with the fuel shut off expansion and compression should appear as one line. Errors would be due to a time lag in the drive or a faulty indicator cam setting or relative phase difference between camshaft and crankshaft. Normally such diagrams would only be necessary on initial engine trials unless loss of compression or cam shift on the engine was suspected.

Fig. 11 is given to show the light spring diagrams for CI engines of the 2- and 4-stroke types. These diagrams are particularly useful in modern practice to give information about the exhaust — scavenge (induction) processes as so many engines utilise turbocharge. The turbocharge effect is shown in each case and it will be observed that there is a general lifting up of the diagram due to the higher pressures.

OTHER RELATED DETAILS

Fuel valve lift cards are very useful to obtain characteristics of injectors when the engine is running. Two diagrams are given as illustration in Fig. 12, relating to Doxford engines although the method is applicable to any type. A standard indicator and drive can be used but a different spring is necessary and the movement from the injector needle is transmitted via a steel probe pin and adaptor. The paper card for examination is laid under an inscribed perspex marking scale to analyse timing characteristics. The early engine illustrates needle characteristics for pilot injection and the modern engine has spring injector lift characteristics. A scavenge pump card is also given in Fig. 12 to show the expected form. Figures given are only generally typical and would vary with the engine type.

Typical diagram faults are normally best considered in the particular area of study where they are likely to occur. However as an introduction two typical combustion faults are illustrated on the draw card of Fig. 12. Turbocharge effects are also shown in Fig. 11 and compression card defects in Fig. 10. It should perhaps be stated that before attempting to analyse possible engine faults it is essential to ensure that the indicator itself and the drive are free from any defect.

Compression ratio has been discussed previously and with SI engines the limits are pre-ignition and detonation. Pinking and its relation to Octane number are important factors as are anti-knock additives such as lead tetra-ethyl Pb $(C_2H_5)_4$. Factors more specific to CI engines are ignition quality, Diesel knock and Octane number, etc. In general these factors plus the important related topics of combustion and the testing and use of lubricants and fuels should be particularly well understood and reference should be made to the appropriate chapter in Volume 8.

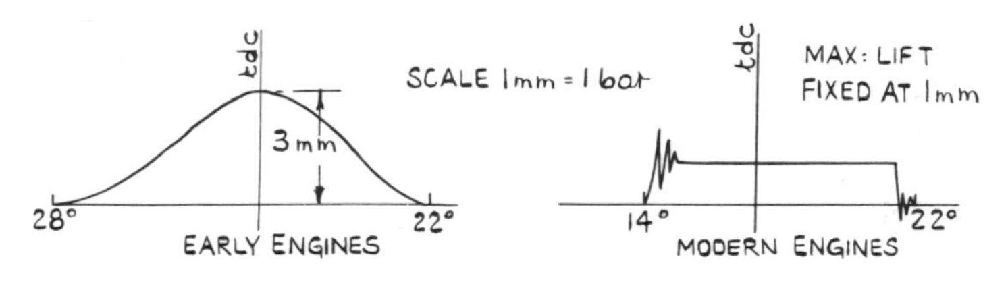

FUEL VALVE LIFT DIAGRAMS (DOXFORD)

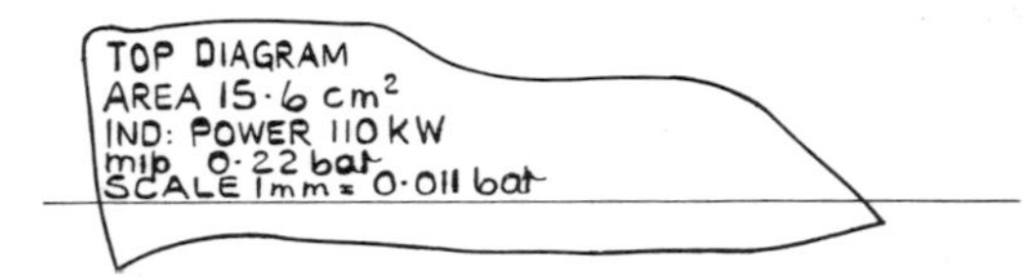

SCAVENGE PUMP CARD

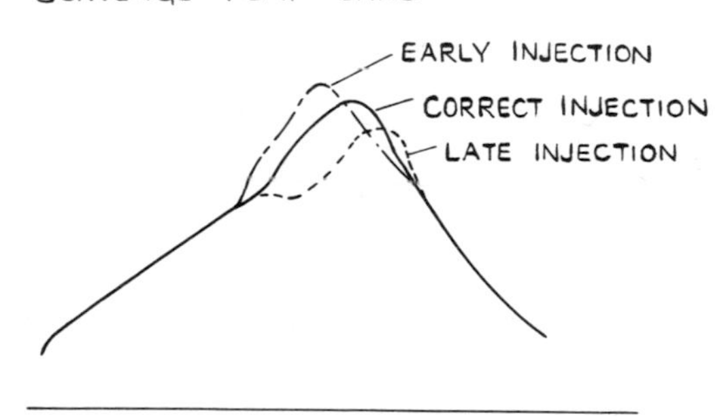

TYPICAL FAULTS SHOWN ON DRAW CARD

RELATED DETAILS.

Fig. 12

Accuracy of indicator diagram calculations is perhaps worthy of specific comment. The area of the power card is quite small and planimeter errors are therefore significant. Multiplication by high spring factors makes errors in evaluation of m.i.p. also significant and certainly of the order of at least ±4%. Further application of engine constants gives indicated power calculations having similar errors. Provided the rather inaccurate nature of the final results is appreciated then the real value of the diagrams can be established. From the power card viewpoint comparison is

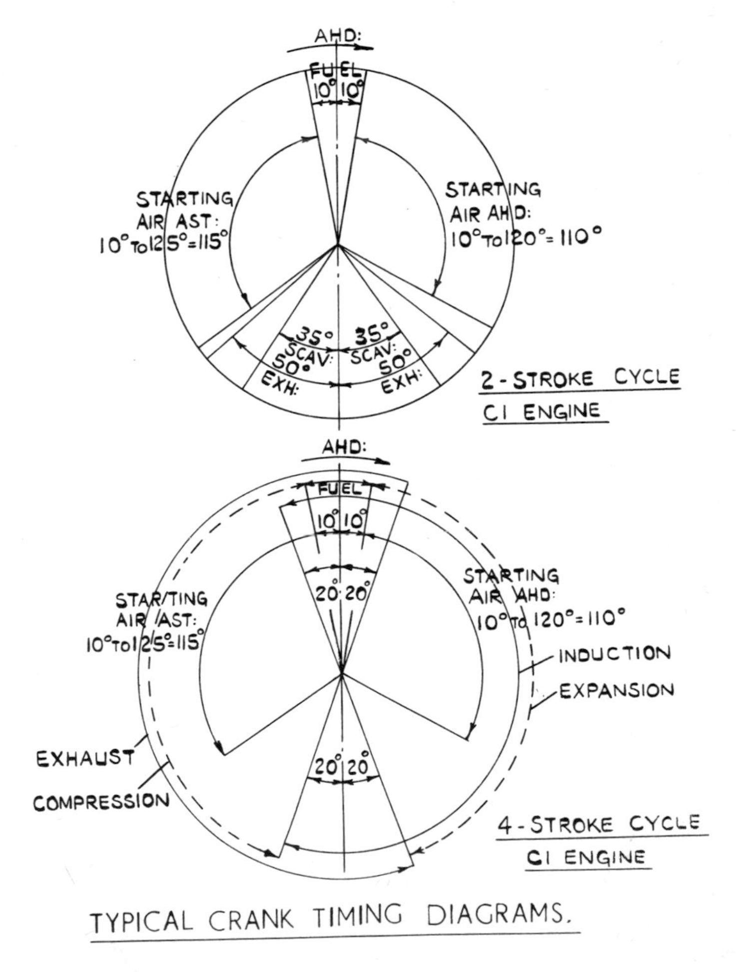

TYPICAL CRANK TIMING DIAGRAMS.

Fig.13

probably the vital factor and indicator diagrams allow this. However modern practice would perhaps favour maximum pressure readings, equal fuel quantities, uniform exhaust temperatures, etc. for cylinder power balance and torsionmeter for engine power. The draw card is particularly useful for compression — combustion fault diagnosis and the light spring diagram for the analysis of scavenge — exhaust considerations.

Crank Timing Diagrams

Typical crank timing diagrams are given in Fig. 13.

2-stroke Cycle

This diagram is for no particular engine but is of a general nature. Fuel injection is fairly typical for modern practice. The starting air period would be satisfactory for 4-cylinder engines and above. Symmetrical scavenge and exhaust periods would be acceptable for non-turbo charged engines. This diagram should be compared to the individual diagrams for specific engine types. The timing diagram is often the basic design starting point once the cycle has been decided and as such is a very important theoretical principle of study in IC engine practice.

4-Stroke Cycle

Again this diagram is of a general nature with symmetrical and simple angular configuration adopted as an aid to memory. Exhaust opening is a little later than is usually found.

Non CI Engines

Sufficient information has been given previously to establish the timing diagram and indicator diagram for a SI engine of the 4-stroke cycle. Hot bulb engine questions can be ignored as obsolete.

Turbocharging

This is considered in detail later in this book but one or two specific comments relating to timing diagrams can be made now. Exhaust requires to be much earlier to drop exhaust pressure quickly before air entry and also requires to be of a longer period to allow discharge of the greater gas mass. Air period is usually slightly greater. This could mean for example in the 2-stroke cycle exhaust from 76 degrees before bottom dead centre to 56 degrees after (unsymmetrical by 10 degrees) and scavenge 40 degrees before and after. For the 4-stroke cycle air open as much as 75 degrees before top centre for 290 degrees and exhaust open 45 degrees before bottom centre for 280 degrees, *i.e.* considerable overlap period.

Actual Timing Diagrams (Doxford)

It is probably useful to consider finally some actual timing diagrams. Four diagrams have been shown on Figs. 14 and 15 to illustrate representative and developing engine design factors with the Doxford engine used for specific example in these cases. Fig. 14 illustrates the early or basic design used for more than thirty

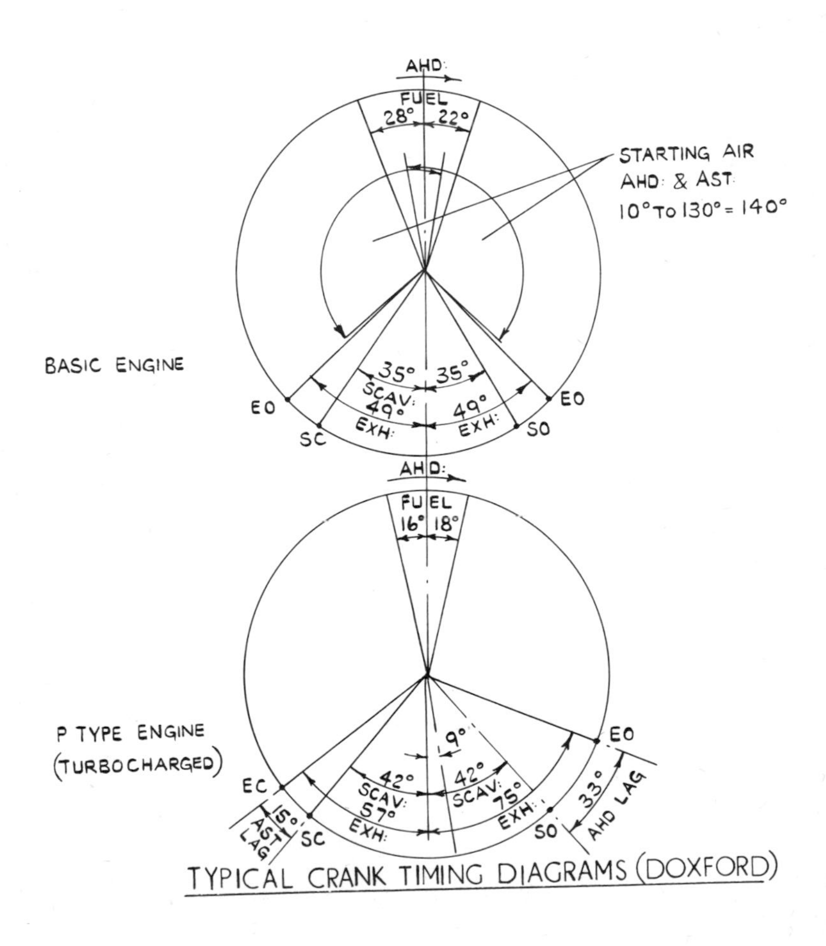

Fig. 14

years. Note air opening before dead centre and the symmetrical form of diagram. As a contrast consider the unsymmetrical timing used on the later P engine design for satisfactory turbocharging. Fig. 15 illustrates an alternative form of diagram often used in modern practice with the P engine used again for example. The unsymmetrical timing created by offset of cranks is clearly seen on this sketch. Fig. 15 completes the development picture by referring to the J engine design of more recent Doxford practice.

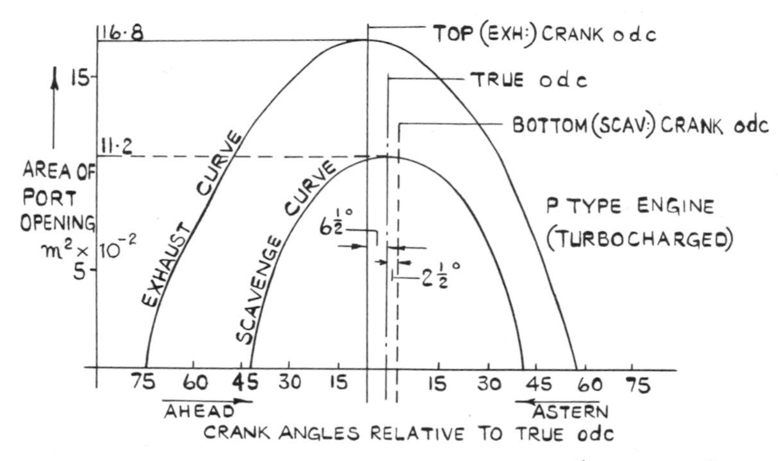

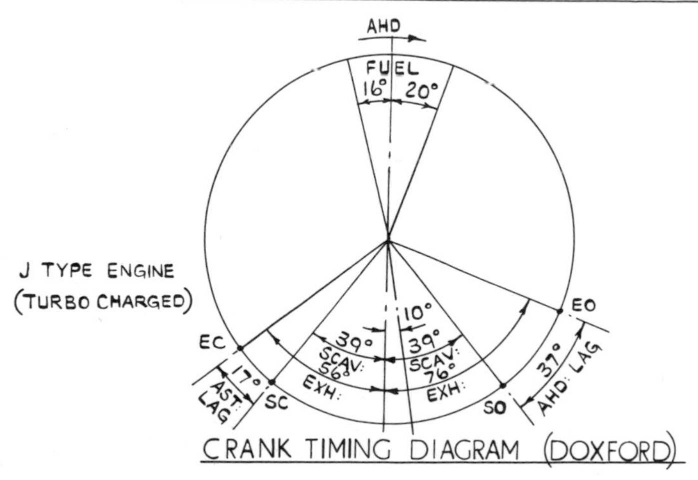

Fig. 15

Types of Indicating Equipment

Conventional indicator gear is fairly well known from practice and manufacturers descriptive literature is readily available for precise details. For high speed engines an indicator of the 'Farnboro' type is often used. Maximum and compression pressures can be taken readily using a peak pressure indicator as sketched in Fig. 16. Counter and adjuster nut are first adjusted so marks on the body coincide at a given pressure on the counter with idler wheel removed. Idler wheel is now replaced. When connected to indicator cock of the engine the adjusting nut is rotated until vibrations of the pointer are damped out. Spring force and gas pressure are now in equilibrium and pressure can be read off directly on the indicating counter (Driven by toothed wheels).

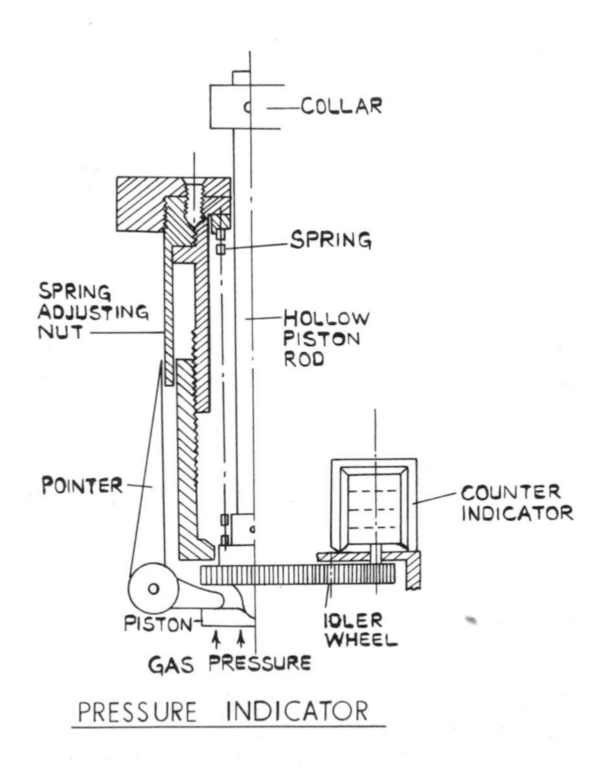

Fig. 16

Modern laboratory practice utilises a cathode ray oscilloscope to give the diagram directly on a screen. By suitable mathematical integration area can be computed and readily converted to m.i.p., indicated power, turning moment, etc. at the turn of the instrument switch. This will obviously find direct application at sea as instrumentation develops. Centralised pressure readings on conventional indicator gear are already fitted by at least one manufacturer of large IC engines.

TEST EXAMPLES 1

Second Class

1. What is meant by the specific fuel consumption per indicated power hour and per brake power hour? Give typical values in each case. How are these affected when running at reduced revolutions as with main engines and also as for a generator with reduced load and constant revolutions?

2. Explain volumetric, scavenge, thermal and mechanical efficiency. Give typical figures in each case stating to which type of engine reference is made. How can these figures be improved?

3. If you were instructed to submit to the superintendent engineer a complete set of indicator diagrams together with the relevant data, give a detailed account of your work in taking off the diagrams and tabulate the data you would forward, giving specimen figures from the last motorship in which you served.

4. Sketch a suitable timing diagram and indicator card for a 2-stroke cycle single acting compression ignition oil engine. State the pressures and temperatures at the main points of the cycle.

5. Describe an indicator suitable for taking indicator cards from a compression ignition oil engine. Describe how a compression card is taken. What is the reason for taking such a card? Sketch a typical compression card.

6. Sketch a valve timing diagram for a 4-stroke CI oil engine. Show the relative positons of the crank. Describe the cycle of operations. Sketch a typical indicator diagram from such an engine and insert average values of temperature and pressure at the cardinal points of such a diagram

First Class

1. Sketch the power, draw and light spring diagrams of the following:—

Show pressures and temperatures of the cardinal points and describe the cycle of events in each case.

(a) 4-stroke naturally aspirated heavy oil engine.

(b) 2-stroke turbocharged heavy oil engine.

(c) 2-stroke heavy oil engine with scavenge blower.

What is the compression pressure and maximum pressure in each case and give reasons why there should be any difference in these figures.

2. Explain, by referring to a theoretical cycle, how the efficiency of an IC engine is dependent on the compression ratio. What criteria determine the compression pressure used in petrol and compression ignition engines?

3. Explain why high compression in the cylinders of an IC engine is considered economical. What determines the maximum compression pressure permissible in SI engines and the minimum compression pressure permissible in CI engines. What would be the result of loss of compression pressure in each case and how would it be rectified?

4. State what is meant by constant volume combustion and constant pressure combustion. Illustrate your answers by diagrams of the cycle and state the pressures used on various engines. Comment on the relative merits of both types in relation to modern engine practice.

5. Sketch a full power indicator diagram with draw card and superimpose the following defects on it:—

(a) Early firing.
(b) Late firing.
(c) After burning.

Explain how an indicator can be used to determine whether or not the camshaft is in correct relationship with the crankshaft.

CHAPTER 2

STRUCTURE AND TRANSMISSION

The engine framework, bedplate and 'A' frames, etc which contain the power transmission system must fulfil the following fundamental requirements and properties:

Strength — is necessary since considerable forces can be exerted. These may be due to out of balance effects, vibrations, gas force transmission and gravitational forces.

Rigidity — is required to maintain correct alignment of the engine running gear. However, a certain degree of flexibility will prevent high stresses that could be caused by slight misalignment.

Lightness — is important, it may enable the power weight ratio to be increased. Less material would be used bringing about a saving in cost. Both are important selling points as they would give increased cargo capacity.

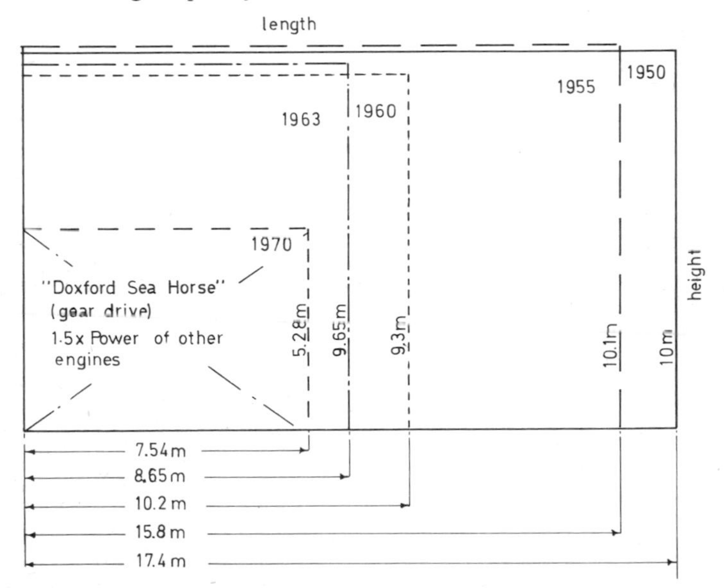

Fig. 17

Toughness — in a material is a measure of its resilience and strength, this property is required to enable the material to withstand the fatigue conditions which prevail.

Simple design — if manufacture and installation are simplified then a saving in cost will be realised.

Access — ease of access to the engine transmission system for inspection and maintenance, and in the first instance installation, is a fundamental requirement.

Dimensions — ideally these should be as small as possible to keep engine containment to a minimum in order to give more engine room space.

Seal — the transmission system container must seal off effectively the oil and vapours from the engine room.

Fig. 17 shows diagrammatically the changes in the principal dimensions that have taken place for one particular engine, but the trend has been the same for all makes. This particular engine manufacturer's latest engine has been added for comparison purposes, but it must be remembered that the higher the speed of operation the smaller can be the engine for the same power.

BEDPLATE

This is generally a one piece structure that may be made of cast iron, prefabricated steel, cast steel, or a hybrid arrangement of cast steel and prefabricated steel.

Cast iron one piece structures are now usually confined to the smaller type of engine, this is because of the cost and weight of the engine. When dealing with castings, problems associated with the mould and pouring increase as the size of the casting increases. Homogeneity of strength, grain size, soundness, impurity segregation, etc all come into play. In addition to the relatively high cost and weight of the cast iron structure, it is not as strong in tension or as stiff as steel. For many years however, cast iron was the preferred material since it gave a stress free, easily machined and at the time a cheap structure.

When welding techniques and methods of inspection improved and larger furnaces became available for annealing, the switch to prefabricated steel structure with its saving in weight and cost was made. It must be remembered that the modulus of elasticity for steel is nearly twice that of cast iron, hence for similar stiffness of structure roughly half the amount of material would be required when using steel.

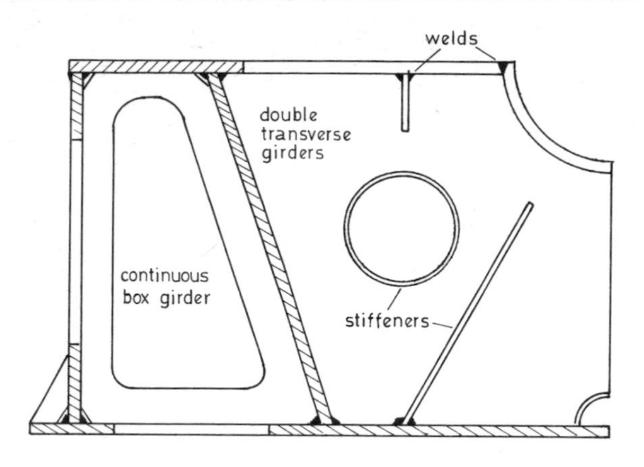

Fig. 18

The Doxford P engine bedplate section Fig. 18 incorporates two longitudinal box girders which extend the full length of the engine to which are welded double transverse girders to carry the main bearing housings. Compared to the earlier types it has increased height and rigidity.

Classification societies only require the transverse girders which are carrying the main bearing housing to be stress relieved. Some manufacturers (*e.g.* Doxford's) stress relieve the complete bedplate, this obviously requires a large annealing furnace. A method to circumvent this is shown in Fig. 19.

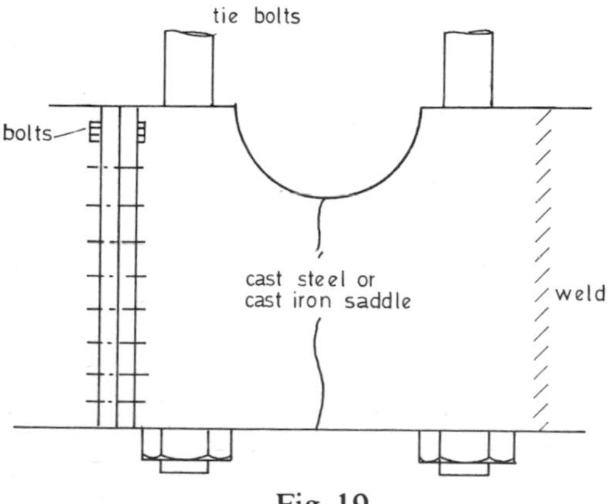

Fig. 19

Fig. 19 shows alternative hybrid arrangements, the main bearing saddle may be cast steel or cast iron. In the former case it would be welded to the prefabricated steel bedplate and in the latter, bolted. These arrangements simplify manufacture and bearing seating alignment.

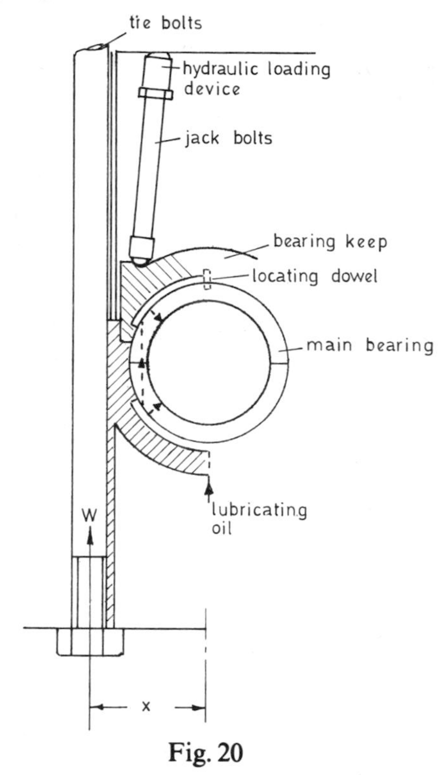

Fig. 20

Structures that have to transmit combustion loading by tie-bolts as well as support the weight of the engine have to be more robust than those used in opposed piston engines. In order to minimise stresses due to bending in the bedplate and hence save some material, the tie-bolts which transmit the combustion forces that cause the bending should be as close as possible to the shaft centre line. Fig. 20 shows diagrammatically the arrangement used in the Sulzer engine. By employing jack bolts, under compression, to retain the bearing keeps in position the distance x is kept to a minimum. Hence the bending moment Wx, where W is the load in the bolt, is also a minimum. The tie rods are pre-stressed at assembly so that welded parts are under compression at all times.

'A' Frames or Columns

These with most engines are prefabricated steel, they carry guide surfaces and are usually bolted to bedplate and cylinder blocks or entablature trunking, the latter being used for air supply purposes jacket and cylinder support, etc.

In the case of a Sulzer engine the 'A' frames are symmetrical and carry oil cooled guides. The cylinder blocks, bedplate and 'A' frames are held together mainly by long through tie bolts and the cylinder blocks are bolted to each other longitudinally.

CRANKSHAFTS

A crankshaft is the backbone of the reciprocating diesel engine, it must be extremely reliable as the cost of replacement would be high, to say nothing of the dangerous situation that may arise on the high seas if it failed.

Types

Fully built, i.e. webs are shrunk on to journals and crank pins.

Semi built, webs and crankpin as one unit shrunk on to the journals.

One piece, one piece of material either cast or forged.

Uses

Large marine diesel engines, fully built or semi-built crankshafts.

Medium speed diesel engines, semi-built.

High speed diesel engines, one piece construction.

In the case of the large marine diesel engine the type of shaft generally favoured is the cast or forged steel semi-built, a typical analysis, method of construction and testing would be as follows.

Material analysis: Cast Steel

Element	*Percentage*
Carbon	0.2
Silicon	0.32
Manganese	0.7
Phosphorus	0.01
Sulphur	0.015
Remainder iron	

Simplified construction

1. Raw material.
2. Raw material refined to remove bulk of the impurities.
3. Metal is degassed in a vacuum furnace to remove hydrogen and nitrogen, etc.
4. Molten metal is then poured into a prepared mould.
5. After removal of the mould the surface of the casting is rough machined, this removes surface and sub-surface defects.
6. Casting is then normalised to improve the grain structure.
7. Stresses are then removed by tempering.
8. Casting is then rough machined to the final dimensions.
9. Crank pin is then cold rolled. By cold rolling the fillet, bending fatigue resistance is increased, micro-defects are reduced and the corrosion fatigue resistance is increased.
10. Finish machining.
11. Web-crank units are shrunk on to the journals.

To ensure that the stress due to shrinkage is not too great, an allowance of 1/550 to 1/700 of the shaft diameter for shrinkage is usual. Reference marks on the outer junction of the crank webs with journals (and crank pins for fully built up types) are provided to give indication of any subsequent slip.

Oil holes in the crankshaft (if any) must be rounded to an even contour with a smooth finish. Fillets should have a radius of not less than 5% of the shaft diameter. Both of the foregoing avoid excessive stress concentration at changes of section.

Tests carried out

1. Chemical analysis, necessary to ensure that the specification is being adhered to and that sulphur and phosphorus contents are not too high.
2. Magnetic particle test, this detects surface and sub-surface defects.
3. Ultra-sonic test, this detects defects within the material.
4. Dye penetrant test, this test detects very slight surface defects.

Tests would also be carried out on material specimens to determine strength, fatigue resistance, etc.

In order to support the large loads imposed by the large bore, high cylinder pressure modern diesel the crankpins and journals must have a large surface area. Crankpin and journal diameters have increased but their lengths reduced in order that overall engine length is a minimum and power-weight ratio is a maximum.

Crankwebs must be kept as thin as possible and to replace strength lost the breadth is increased. In some engines the crank-web has become the main journal bearing thus greatly reducing engine length.

Balance weights are sometimes fitted to crank webs to reduce out of balance forces and couples. Out of balance effects which vibrate the engine, can lead to a variety of defects.

It has already been stated that a crankshaft must be extremely reliable, if we examine the stresses to which a crankshaft is subjected then we may appreciate the need for extreme reliability.

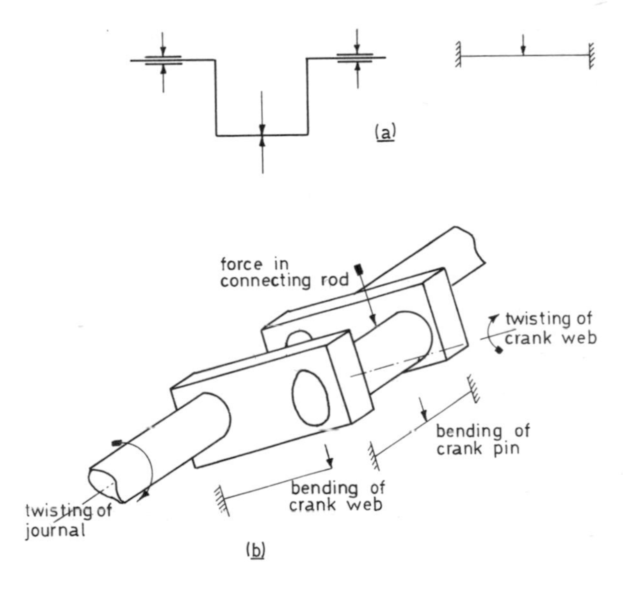

STRESSES IN CRANKSHAFT

Fig. 21

Fig. 21 shows a crank unit with equivalent beam systems. Diagram (a) indicates the general, central variable loaded, built-in beam characteristic of a crank throw supported by two main bearings. If the bearings were flexible. *e.g.* spherical or ball, then a simply supported beam equivalent would be the overall characteristic.

Examining the crank throw in greater detail, diagram (b), shows that the crank pin itself is like a built-in beam with a distributed load along its length that varies with crank position. Each crank web is like a cantilever beam subjected to bending and twisting. Journals would be principally subjected to twisting, but a bending stress must also be present if we refer back to diagram (a).

Bending causes tensile, compressive and shear stresses. Twisting causes shear stress.

Due to shrinkage of the crankwebs onto the journals (and in the fully built up case also the crank pins) compressive stresses are set up in the journals and tensile hoop stresses in the crank webs.

In heavy weather conditions the stresses set up in the various parts of the crankshaft mentioned above will be accentuated. Torsional stress due to power transmission is going to fluctuate widely if the propeller comes out of the water and then plunges back. Sudden fluctuations of engine speed result in shock loading on crank pins (this is caused by inertia forces due to large accelerations of reciprocating masses). Pounding of propeller blades into the water can lead to impulsive forces causing vibration of the shafting which will further increase the magnitude of the stresses especially if the pounding frequency is at or near the natural frequency.

Crankshaft defects and their causes

Misalignment

If we assume that alignment was correct at initial assembly then possible reasons for misalignment are as follows:—

1. Worn main bearings. Caused by incorrect bearing adjustment leading to overloading. Broken, badly connected or choked lubricating oil supply pipes causing lubrication starvation. Contaminated lubricating oil. Vibration forces.

2. Excessive bending of engine framework. This could be caused by incorrect cargo distribution but is unlikely, more probable that the cause would be grounding of the vessel, it being re-floated in a damaged condition. It is essential that all bearing clearances be checked and crankshaft deflections taken after such an accident.

Vibration

This can be caused by: incorrect power balance, prolonged running at or near critical speeds, slipped crank webs on journals, light ship conditions leading to impulsive forces from the propeller (*e.g.* forcing frequency four times the revs. for a four-bladed propeller), the near presence of running machinery, excessive wear down of the propeller shaft bearing (this in bad weather conditions can lead to whipping of the shafting).

Vibration accentuated stresses, they can be increased to exceed fatigue limits and considerable damage could result. It can lead to things working loose, *e.g.* coupling bolts, bearing bolts, bolts securing balance masses to crank webs and lubricating oil pipes.

Other causes

Incorrect manufacture leading to defects is fortunately a rare occurrence. In the past, failure has been caused by: slag inclusions, heat treatment and machining defects (*e.g.* badly radiused oil holes and fillets).

Crank webs have slipped on journals; this could be caused by seizure of some component, *e.g.* bearing, guide shoe, piston rod in gland or piston in liner. Bottom end bearing bolt failure, or starting the engine with the turning gear in, have also caused slip due to shock load.

A few words about bottom end bearing bolt failure would be appropriate at this point. This type of bolt failure, especially in high speed auxiliary diesel engines, occurred relatively frequently in the 1940 to 1960 era. The main cause being exceeding fatigue limit stresses.

Reasons for exceeding stresses are various, *e.g.* overtightening of bolts (especially those with castellated nuts), time in service too long and bad design.

In the event of bottom end bearing bolt failure, considerable damage may result, it depends upon the position of the failure. If breakage was just below the nut then the bolt would not be able to fall out of the bearing. It would then tend to plough its way around the bottom of the crank pit, bearing failure would probably result and the remaining bolt may fail. It is then possible that the top half of the bearing may leave the crank pin and be struck by the crank pin. This could then lead to piston, cylinder cover, cylinder liner, jacket, piston rod and connecting rod damage.

Such an event occurred within the author's experience and in addition to all the damage outlined above, the crankshaft of the engine was badly scored but fortunately only slightly twisted. Repairs consisted of dressing up the crankshaft, metal-locking the crankcase, the crank pit and the jacket. Renewal of liner, cylinder cover, valves, piston, piston rod, connecting rod, bottom end bearing and one main bearing.

Fretting corrosion

Occurs where two surfaces forming part of a machine, which in theory constitute a single unit, undergo slight oscillatory motion of a microscopic nature.

It is believed that the small relative motion causes removal of metal and protective oxide film. The removed metal combines with oxygen to form a metal oxide powder that may be harder than the metal (certainly in the case of ferrous metals) thus increasing the wear. Removed oxide film would be repeatedly replaced, increasing further the amount of damage being done.

Fretting damage increased with load, amplitude of movement and frequency. Hardness of the metal also effects the attack, in general damage to ferrous surfaces is found to decrease as hardness increases.

Oxygen availability also contributes to the attack, if oxygen level is low the metal oxides formed may be softer than the parent metal thus minimising the damage. Moisture tends to decrease the attack.

Bearing Corrosion

In the event of fuel oil and lubricating oil combining in the crankcase, weak acids may be released which can lead to corrosion of copper lead bearings. The lead is removed from the bearing surface so that the shaft runs on nearly pure copper, this raises bearing temperature so that lead rises to the surface and is removed. The process is repeated until failure of the bearing takes place. Scoring of crankshaft pins can then occur. Use of detergent types of lubricating oil can prevent or minimise this type of corrosion. The additives used in the oil to give it detergent properties would be alkaline, in order to neutralise the weak acids.

Water in the lubricating oil can lead to white metal attack and the formation of a very hard black incrustation of tin oxide. This oxide may cause damage to the journal or crankpin surface by grinding action.

Bearing Clearances and Shaft Misalignment

Bearing clearances can be checked in a variety of ways, a rough check is to observe the discharge of oil, in the warm condition, from the ends of the bearings. Feeler gauges can be used, but for some of the bearings they can be difficult to manoeuvre into position in order to obtain readings. Clock (or as they are sometimes called, dial) gauges can be very effective and accurate providing the necessary relative movement can be achieved, this can prove to be very difficult in the larger types of engine. Finally, the use of lead wire necessitating the removal of the bearing keeps.

Main bearing clearances, should be zero at the bottom. If they are not, then the crankshaft is out of alignment. Some engines are provided with facilities for obtaining the bottom clearance (if any) of the main bearings with the aid of special feelers, without the need to remove the bearing keep. Another method is to first arrange in the vertical position a clock gauge so that it can record the movement of the crank web adjacent to the main bearing. The main bearing keep is then removed, shims are withdrawn and the keep is replaced and tightened down. The vertical movement of the shaft, if any, is observed on the dial gauge.

Obviously, if the main bearing clearance is not zero at the bottom the adjacent bearing or bearings are high by comparison and the shaft is out of alignment.

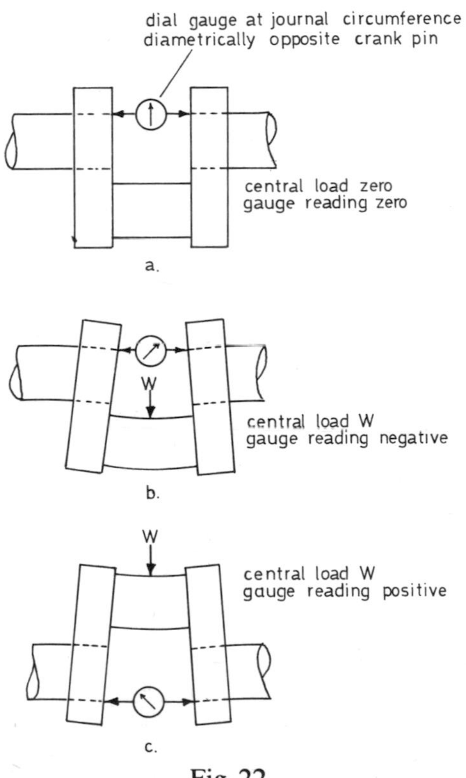

Fig. 22

Crankshaft alignment can be checked by taking deflections. If a crank throw supported on two main bearings is considered, the vertical deflection of the throw in mid span is dependent upon: shaft diameter, distance between the main bearings, type of main bearing, and the central load due to the running gear. A clock gauge arranged horizontally between the crank webs opposite the crank pin and ideally at the circumference of the main journal (see Fig. 22) will give a horizontal deflection, when the crank is rotated through one revolution, that is directly proportional to the vertical deflection.

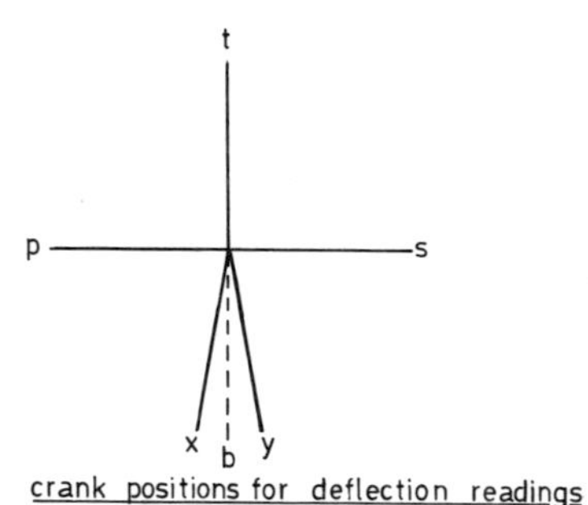

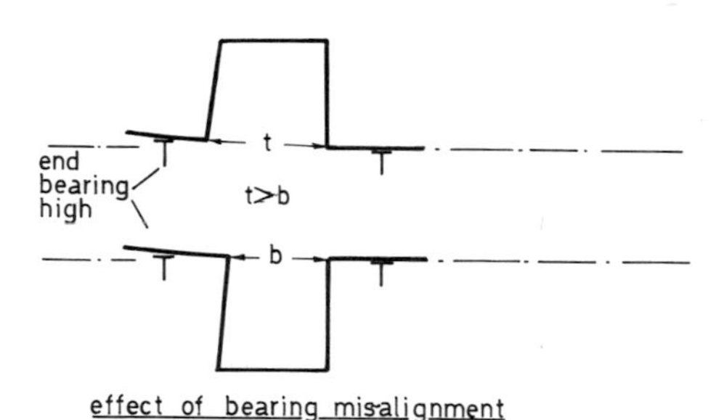

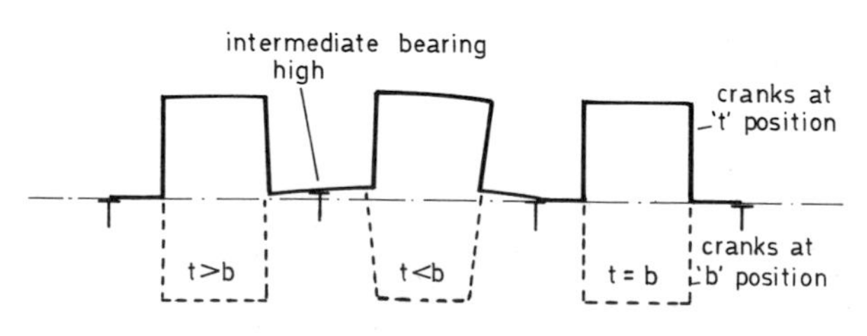

Fig. 23

In Fig 22 (a) it is assumed that main bearings are in correct alignment and no central load is acting due to running gear, then vertical deflection of the shaft would be small — say zero. With running gear in place and crank at about bottom centre the webs would close in on the gauge as shown — this is negative deflection. With crank on top centre webs open on the gauge — this is positive deflection.

In practice the gauge must always be set up in the same position between the webs each time, otherwise widely different readings will be obtained for similar conditions. An alternative is to make a proportional allowance based on distance from crankshaft centre. Obviously the greater the distance from the crankshaft centre the greater will be the difference in gauge readings between bottom and top centre positions.

Since, due to the connecting rod, it is generally not possible to have the gauge diametrically opposite crank pin centre when the crank is on bottom centre an average of two readings would be taken, one either side during the turning of the crank.

The following table shows some possible results from a six cylinder diesel engine:

GAUGE READINGS IN mm/100

	CYLINDER NUMBER					
CRANK POSITION	1	2	3	4	5	6
x	0	0	0	0	0	0
p	5	2	6	–8	–3	1
t	10	3	12	–14	–8	4
s	5	3	6	–8	–6	3
y	–2	2	–2	0	0	–2
b = (x + y)/2	–1	1	–1	0	0	–1
Vertical mis-alignment (t-b)	11	2	13	–14	–8	5
Horizontal mis-alignment (p-s)	0	–1	0	0	3	–2

The dial gauge would be set at zero when crank is in, say, port side near bottom position and gauge readings would be taken at port horizontal, top centre, starboard horizontal and starboard side near bottom positions. Say x, p, t, s and y as per Fig 23, but before taking each reading the turning gear should be reversed to unload the gear teeth, otherwise misleading readings may be obtained.

Engines with spherical main bearings will have greater allowances for crankshaft misalignment than those without. Spherical bearings are used when increased flexibility is required for the crankshaft. This would be the case for opposed piston engines with large distances between the main bearings, so instead of having a built-in beam effect the arrangement is more likened to a simply supported beam, with its larger central deflection for a given load.

From the vertical misalignment figures and by referring to Fig. 23 the reader should be able to deduce that, the end main bearing adjacent to No. 1 cylinder and the main bearing between Nos. 3 and 4 cylinders are high.

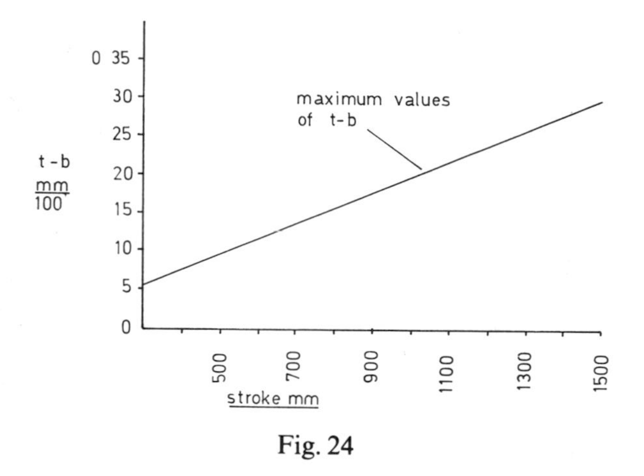

Fig. 24

Vertical and horizontal misalignment can be checked against the permissible values supplied by the engine builder, often in the form of a graph as per Fig. 24. If any values exceed or equal maximum permissible values then bearings will have to be adjusted or renewed where required. Indication of incorrect bearing clearances may be given when the engine is running. In the case of medium or high speed diesels, single acting, and large slow running diesels double acting, load reversal at the bearings generally occurs. With excessive bearing clearances loud knocking takes place, white metal then usually gets hammered out.

If bearing lubrication, for a unit, is from the same source as piston cooling, then a decrease in the amount of cooling oil return, may be observed in the sight glass, together with an increase in its temperature.

If bearing clearances are too small, overheating and possible seizure may take place. Oil mist and vapour at a particular unit may be observed to increase — together with a hot bearing, this may lead to a crankcase explosion.

Regular checks must be made to ascertain the oxidation rate of the oil. If this is increasing then high temperatures are being encountered. n.b. as the oil oxidises (burns) its colour blackens.

CHOICE, MAINTENANCE AND TESTING OF LUBRICATING OIL FOR MAIN CIRCULATING SYSTEM

Choice

If the engine is a 'trunk type' then fuel and deleterious deposits from its combustion products may find their way into the crankcase. The oil should therefore be one which has detergent properties, these oils are sometimes called 'Heavy Duty'. Additives in these oils deter the formation of deposits by keeping substances, such as carbon particles, in suspension. They also counteract the corrosive effect of sulphur compounds, some of the fuels used may be low in sulphur content, in this case the alkaline additive in the lubricating oil could be less.

Detergent oils may not be able to be water washed in a centrifuge, it is always advisable to consult the supplier.

Straight mineral oil, generally with an anti-oxidant and corrosion inhibitor added, is the type normally used in diesels whose working cylinder is separate from the crankcase.

Maintenance

When the engine is new correct pre-commissioning should give a clean system free from sand, metal, dust, water and other foreign matter. To clear the system of contaminants all parts must be vibrated by hammering or some other such method to loosen rust flakes, scale and weld spatter (if this is not done then these things will work loose when the engine is running and cause damage). A good flushing oil should then be used and clear discharges obtained from pipes before they are connected up, filters must be opened up and cleaned during this stage. Finally, the flushing operation should be frequently repeated with a new charge of oil of the type to be used in the engine.

When the engine is running, continuous filtration and centrifugal purification are necessary. Centrifugal purification, on the continuous by-pass principle, may have to be accompanied by a water wash from the very start in order to keep the oil in good condition. In the case of detergent type oils the water wash may be limited or not allowed since the water may remove some of the additives.

Oxidation of the oil is one of the major causes of its deterioration, it is caused by high temperatures. These may be due to:

1. Small bearing clearances (hence insufficient cooling).

2. Not continuing to circulate the oil upon stopping the engine. In the case of oil cooled piston types, piston temperatures could rise and the static oil within them become overheated.

3. Incorrect use of oil preheater for the purifier, *e.g.* shutting off oil before the heat or running the unit part full.

4. Metal particles of iron and copper can act as catalysts that assist in accelerating oxidation action. Rust and varnish products can behave in a similar fashion.

When warm oil is standing in a tank, water that may be in it can evaporate and condense out upon the upper cooler surfaces of the tank not covered by oil. Rusting could take place and vibration may cause this rust to fall into the oil. Tanks should be given some protective type of coating to avoid rusting.

Drainings from scavenge spaces and stuffing boxes should not be put into the oil system and stuffing box and telescopic pipe glands must be maintained in good condition to prevent entry of water, fuel and air into the oil system.

Regular examination and testing of the main circulating oil is important. Samples should be taken from a pipeline in which the oil is flowing and not from some tank or container in which the oil is stationary and could possibly be stagnant.

Smelling the oil sample may give indication of fuel oil contamination or if acrid, heavy oxidation. Dark colour gives indication of oil deterioration, due mainly to oxidation.

Dipping fingers into the oil and rubbing the tips together can detect reduction in oiliness — generally due to fuel contamination — and the presence of abrasive particles. The latter may occur if a filter has been incorrectly assembled, damaged or automatically by-passed. Water vapour can condense on the surfaces of sight glasses, thus giving indication of water contamination. But various tests are available to detect water in oil, *e.g.* immersing a piece of glass in the oil, water finding paper or paste — copper sulphate crystals change colour from white to blue in the presence of water — plunging a piece of heated metal such as a soldering iron into the oil causes spluttering if water is present.

A check on the amount of sludge being removed from the oil in the purifier is important, an increase would give indication of oil deterioration. Lacquer formation on bearings and excessive carbon formation in oil cooled pistons are other indications of oil deterioration.

Oil samples for analysis ashore should be taken about every 3000 hours (or more often if suspect) and it would be recommended that the oil be changed if one or more of the following limiting values are reached:

1. 5% change in the viscosity from new. Viscosity increases with oxidation and by contamination with heavy fuel, diesel oil can reduce viscosity.

2. 0.5% contamination of the oil.

3. 0.5% emulsification of the oil, this is also an indication of water content. Water is generally permissible up to 0.2%, dangerous if sea water.

4. 1.0% Conradson carbon value. This is from cracked lubricating oil or residue from incomplete combustion of fuel oil.

5. 0.01 mg KOH/g Total Acid Number (TAN). The TAN is the total inorganic and organic acid content of the oil. Sulphuric acid from engine cylinders and chlorides from sea water give the inorganic, oxidation produces the weak organic acids. Sometimes the acids may be referred to as Strong and Weak.

LUBRICATION SYSTEMS

Lubrication systems for bearing and guides, etc. should be simple and effective. If we consider the lubrication of a bottom end bearing, various routes are available, the object would be to choose that route which will be the most reliable, least expensive and least complicated. We could supply the oil to the main bearing and by means of holes drilled in the crankshaft convey the oil to the bottom end bearing. This method may be simple and satisfactory for a small engine but with a large diesel it presents machining and stress problems.

In one large type of diesel the journals and crankpins were drilled axially and radially, but to avoid drilling through the crank-web and the *shrinkage surfaces* the oil was conveyed from the journal to the crank pin by pipes.

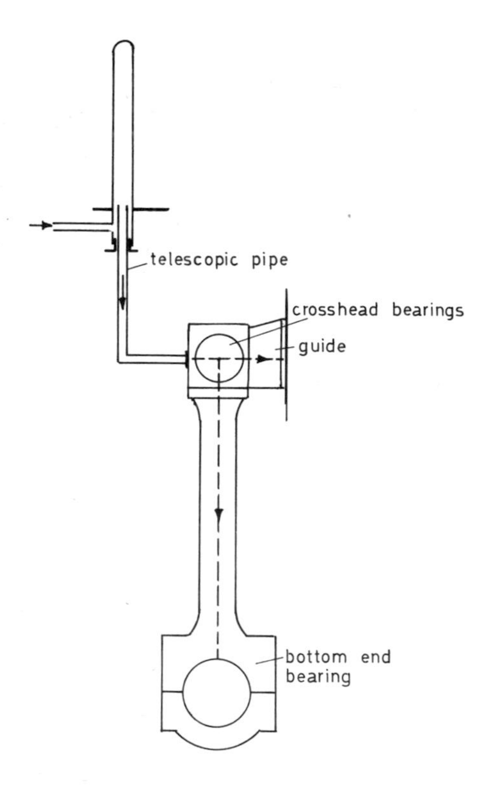

LUBRICATION OF TOP, BOTTOM & GUIDE BEARINGS THROUGH TELESCOPIC SYSTEM

Fig. 25

A common arrangement, mainly adopted with engines having oil cooled pistons, is to supply the bottom end bearing with oil down a central hole in the connecting rod from the top end bearing. Fig. 25 shows an arrangement wherein a telescopic pipe system is used and Fig. 26 a swinging arm, the disadvantage of the latter is that it has three glands whereas the telescopic has only one. However, it is more direct and could be less expensive.

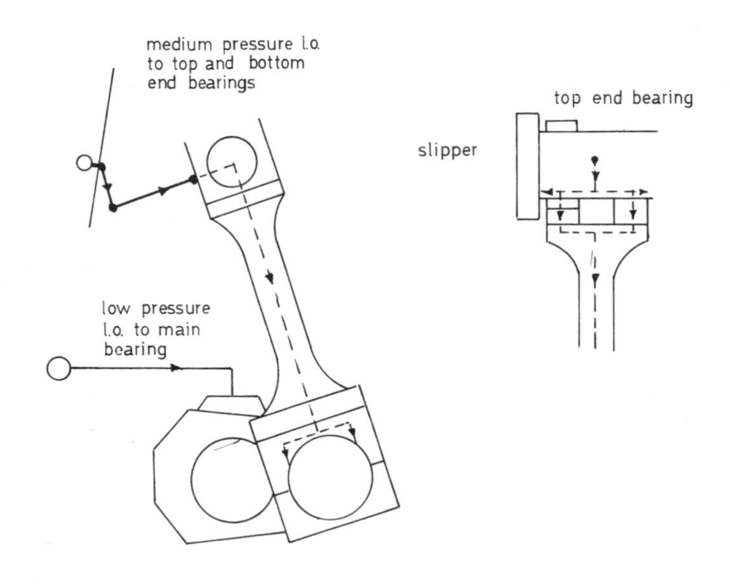

R.N.D. LUBRICATION SYSTEM FOR
MAIN BOTTOM AND TOP END BEARINGS

Fig. 26

With any of the bearings (excepting ball or roller) the main object is to provide as far as possible a good hydrodynamic film of lubricant (*i.e.* a continuous unbroken film of oil separating the working surfaces). Those factors assisting hydrodynamic lubrication are:—

1. *Viscosity.* If the oil viscosity is increased there is less liklihood of oil film break down. However, too high a viscosity increases viscous drag and power loss.

2. *Speed.* Increasing the relative speed between the lubricated surfaces pumps oil into the clearance space more rapidly and helps promote hydrodynamic lubrication.

3. *Pressure.* Increasing bearing load and hence pressure (load/area) breaks down the oil film. In design, if the load is increased area can be increased by making the pin diameter larger — this will also increase relative speed.

4. *Clearance.* If bearing clearance is too great inertia forces lead to 'bearing knock.' This impulsive loading results in pressure above normal and breakdown of the hydrodynamic layer. Fig. 27 illustrates the foregoing points graphically for a journal type of bearing.

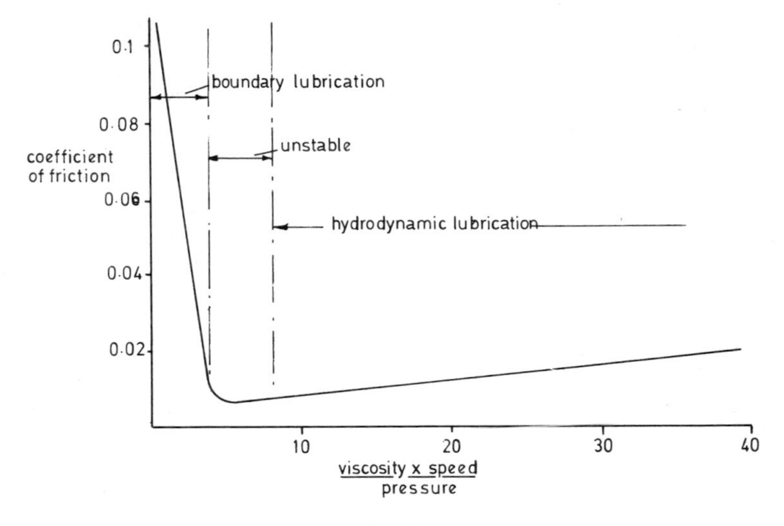

Fig. 27

Hydrodynamic lubrication should exist in main, bottom end and guide bearings. The top end bearing will have a variable condition, *e.g.* when at T.D.C. relative velocity between crosshead pin and bearing surface is zero and bearing pressure near or at maximum. Methods of improving top end bearing lubrication are:—

1. Reversal of load on top end by inertia forces — only possible with medium or high speed diesels.

2. Use as large a surface area as possible, *i.e.* the complete underside of the crosshead pin.

3. Avoid large axial variation of bearing pressure by more flexible seating and design.

4. Increase oil supply pressure. Fig. 28 shows a method of increasing oil supply pressure to the top end bearing which tends to keep the crosshead pin 'floating' at all times. As the connecting rod oscillates, lubricating cross head oil is pumped at high pressure from the two pumps (only one is shown).

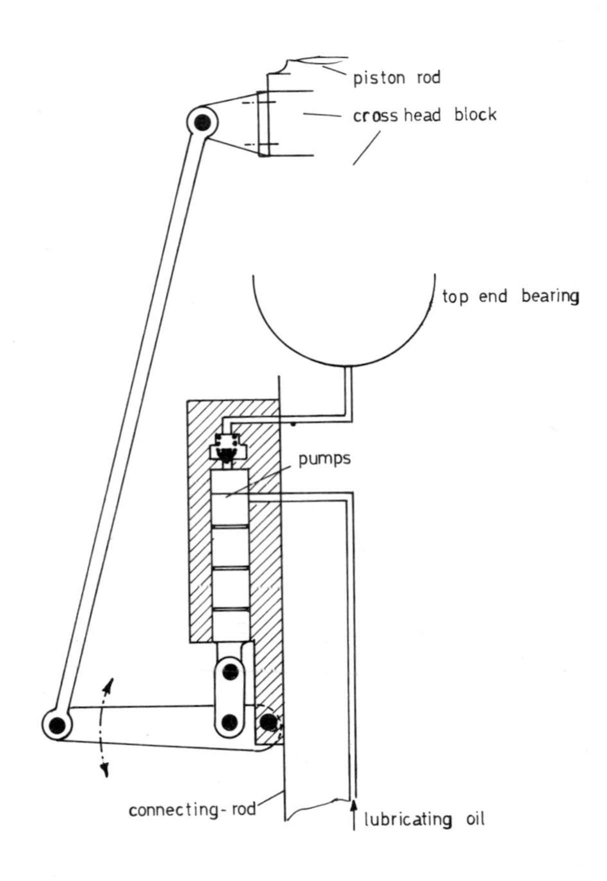

Fig. 28

Doxford crosshead

Fig. 29 is a simplified sketch suitable for examination purposes and it shows the method used to supply lubricating oil for top end, bottom end and guide bearing lubrication together with piston cooling. Oil supply and return is by telescopic pipes.

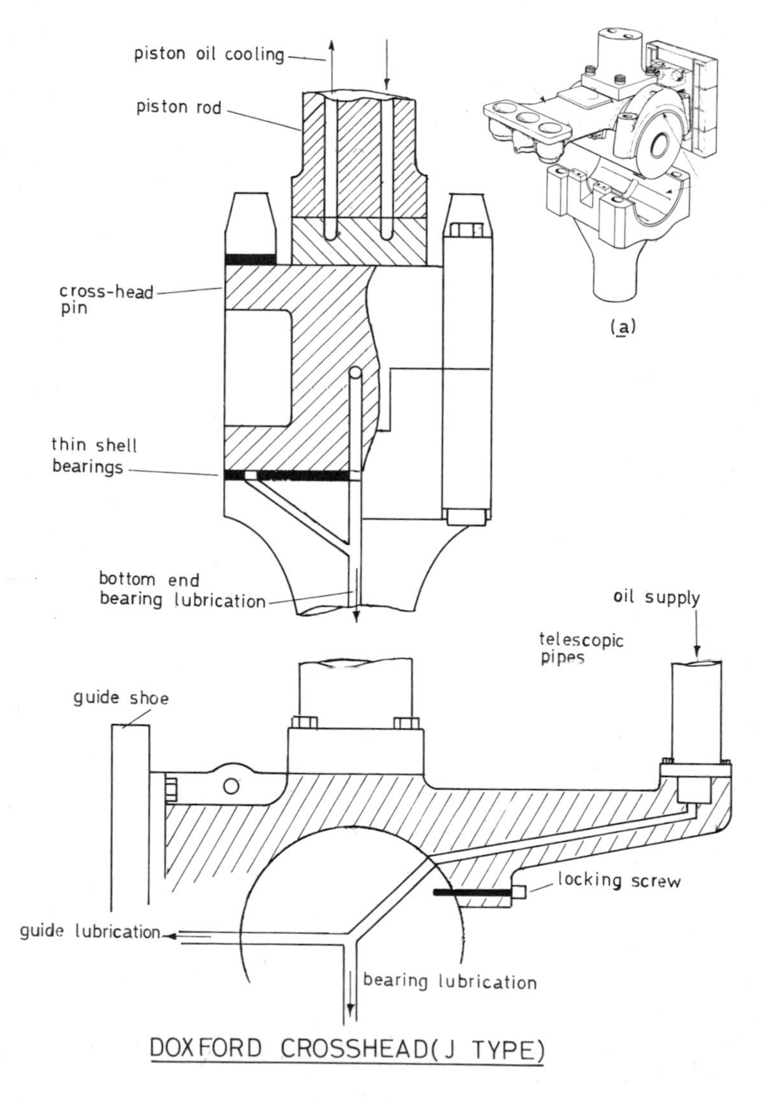

Fig. 29

The whole underside of the locked crosshead pin is supported on a bearing surface thus minimising bearing pressure and encouraging hydrodynamic lubrication. Oil supply is via the circumferential and longitudinal oil grooves, the inset Fig. 29 (a) shows the arrangement clearly.

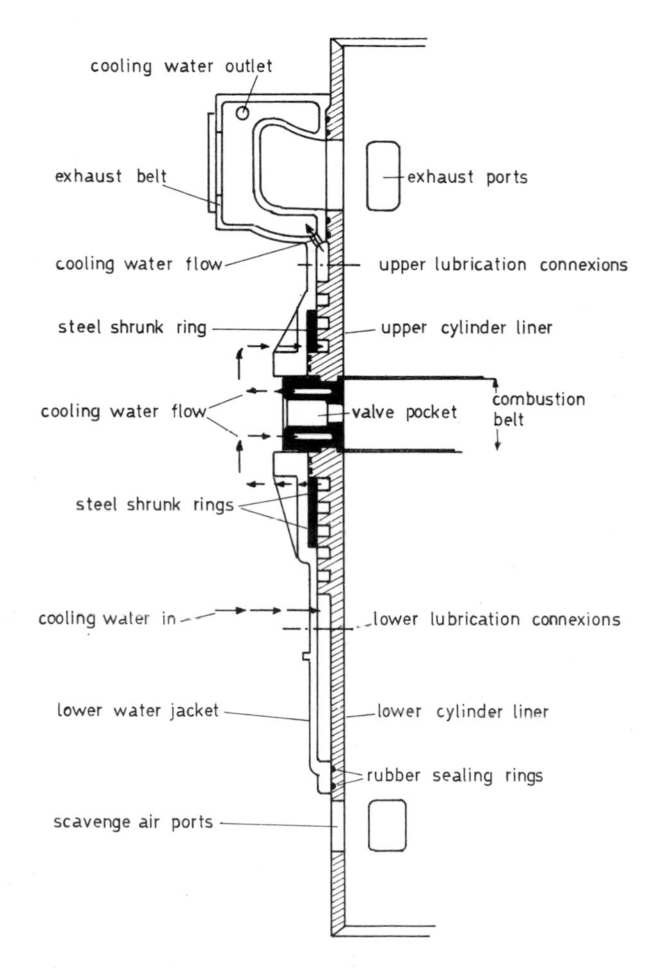

TRIPARTITE CYLINDER LINER & JACKET

Fig. 30

CYLINDERS AND PISTONS

Cylinders

Many different cylinder arrangements exist, but as there are common practical factors for all only one will be described in any detail.

Fig. 30 shows in section a modern tripartite (*i.e.* three part) cylinder liner and jacket of the type used in the Doxford P engine. When sketching such an arrangement for examination purposes draw only one half, as shown, since the liner is symmetrical.

Upper and lower liners are made thin to give good heat transfer and have supporting ribs for strength and cooling water passage. Steel rings are shrunk on to the supporting ribs to give additional strength to withstand combustion loads. The lower water jacket, exhaust and combustion belts are bolted together. This clamps the two wearing parts of the cylinder liner to the valve carrying combustion belt. Main advantages of such a system are:—

1. Liners are simple, short, hard wearing iron castings which can be relatively easily manufactured.

2. Each half (top or bottom) of the liners can be separately replaced if required.

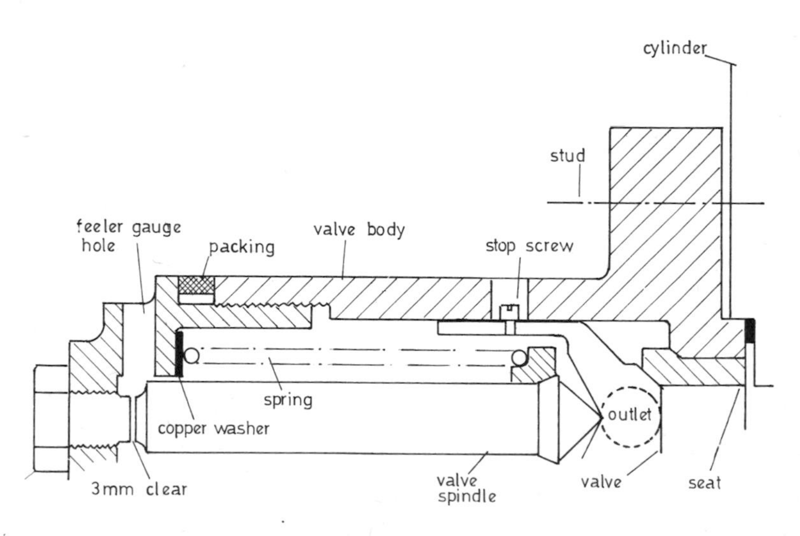

Fig. 31

3. Strong cast steel can be used for the combustion belt whose wearing properties does not have to match up to the cylinder liners.

A main disadvantage is the spigoted copper joints, with some arrangements of cylinders there are no joints, or possibly one.

Expansion longitudinally of the cylinder liner is allowed for through the exhaust belt at the upper end and through the lower water jacket at the bottom end, water sealing is accomplished by rubber rings across which the liner moves as it expands.

The combustion belt has valve openings for two fuel, one starting and one relief valve which are fitted to jointed flanges directly on to the combustion belt, no glands or packing is required.

A typical relief valve is shown in Fig. 31, the special heat resisting steel valve is limited to 3 mm opening by the screw at the end of the body and is set to lift and relieve pressures in excess of about 65.5 bar, *i.e.* about 10% above maximum pressure. Rotation of the valve is prevented by the stop screw.

Cylinder liner wear

With correct cylinder lubrication, correctly fitted piston rings and warming through of the diesel engine before starting together with good combustion and properly timed fuel injection, cylinder liner wear should be kept to a minimum. However, with the advent of heavy fuel burning, which gives the diesel engine considerable economic advantages over steam turbines, increased wear rate in cylinders and increased piston ring breakage occurred.

Wear rates of up to 0.8mm/1000h have been experienced and excessive piston ring breakage has been caused by the variable stresses set up due to the opening and closing as the ring passes from a region of high wear to that of low wear. Pistons have also failed due to thermal fatigue.

Some of the more modern metals used for cylinder liners (and more expensive), *e.g.* Vanadium-titanium cast iron has reduced the wear rate to some 0.3mm/1000h and less. Another method of keeping cylinder wear rates down is to use chromium plates liners (up to three times more wear is possible on a unplated liner of the pearlitic iron type).

Chromium plating of cylinder liners results in a non porous surface, this means that the lubricating oil would not be retained. In order that the oil can be retained, porosity is manufactured. Two ways of doing this are: (1) immersion of the plated cylinder liner into a special bath and reversal of current, this forms minute pits and channels in the chromium, (2) after removal of the liner from the plating bath it is ground or honed and then replaced in the bath together with a screen attached to its surface. The current is then reversed and small hemispherical cavities are produced.

Cylinder Lubrication

The principal objects of cylinder lubrication are:—

1. To separate sliding surfaces with an unbroken oil film.

2. To form an effective seal between piston rings and cylinder liner surface to prevent blow past of gases.

3. To neutralise corrosive combustion products and thus protect cylinder liner, piston and rings from corrosive attack.

4. To soften deposits and thus prevent wear due to abrasion.

5. To remove, dissipate and cause the loss of deposits to exhaust, hence preventing seizure of piston rings and keeping engine clean.

6. To cool hot surfaces without burning.

In practice some oil burning will take place, if excessive this would be indicated by blue smoke and increased oil consumption. As the oil burns it should leave as little and as soft a desposit as possible. Over lubrication should be avoided.

When the engine is new, cylinder lubrication rate should normally be greater than when the engine is run in. Reasons for this increased lubrication are: (1) surface asperities will, due to high local temperatures, cause increased oxidation of the oil and reduce its lubrication properties, (2) sealing of the rough surfaces is more difficult, (3) worn off metal needs to be washed away.

The actual amount of lubricating oil to be delivered into a cylinder per unit time depends upon: stroke, bore and speed of engine, engine load, cylinder temperature, type of engine, position of cylinder lubricators and type of fuel being burnt.

Position of the cylinder lubricators for injection of oil has always been a topic of discussion, the following points are of importance.

1. They must not be situated too near the ports, oil can be scraped over edge of ports and blown away.

2. They should not be situated too near the high temperature zone or the oil will burn easily.

3. There must be sufficient points to ensure as even and as complete a coverage as possible.

Ideally, timed injection of lubricant delivering the correct measured quantity to a specific surface area at the correct time in the cycle is the aim. Fig. 32 shows a typical cylinder lubricator for a modern diesel engine and how it can be removed without disturbing the water seals.

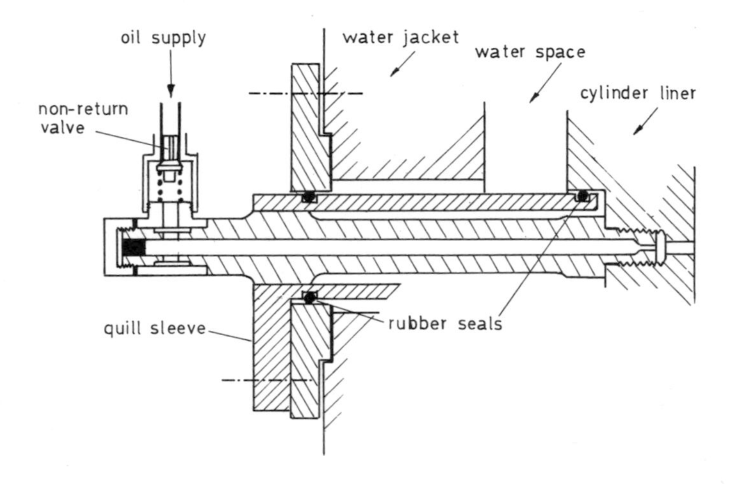

CYLINDER LUBRICATOR

Fig. 32

PISTONS AND RINGS

Pistons

Various materials have been used and are used for pistons. Cast iron (today this would be spheroidal graphitic), forged steel and cast steel are materials favoured for the piston crowns of large marine diesel engines.

Fig. 33 shows a cast steel piston for a large Sulzer engine. The material combines strength and wear resistance, the thickness can be kept to a minimum improving heat transfer and minimising risk of thermal cracking or distortion. Where additional strength is required thin strengthening ribs are provided.

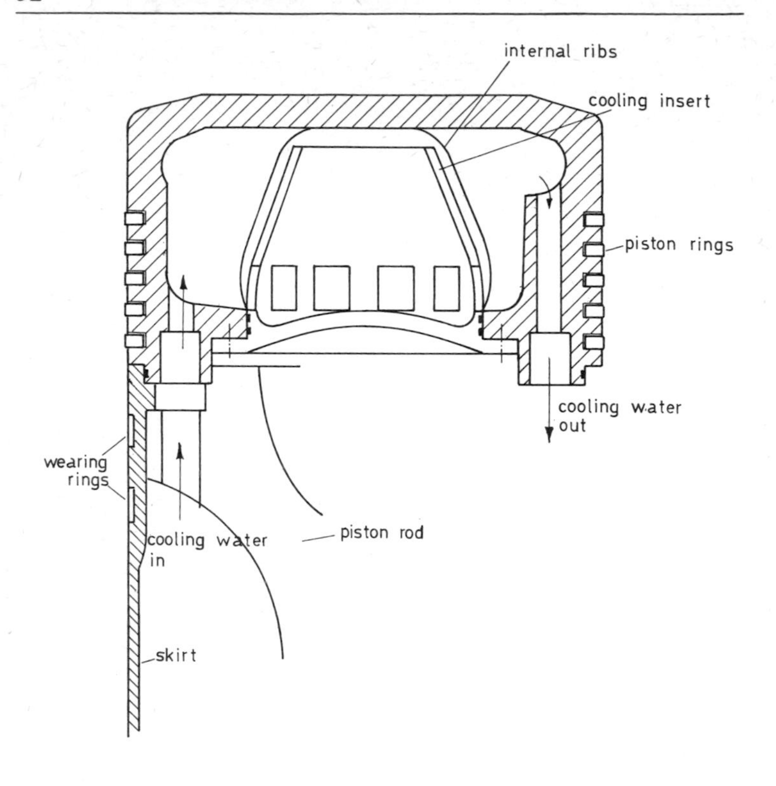

WATER COOLED PISTON

Fig. 33

Intensive cooling of this piston is achieved by the cocktail shaker effect of the water. With air present in the piston (this comes from the telescopic system, it being necessary to provide a cushion and prevent water hammer) together with water, the inertia effect coupled with the conically shaped insert leads to very effective cooling by secondary flow as the crank goes over t.d.c.

It is claimed that metal temperature in the hottest region is reduced by 40°C. It must be borne in mind that at slow speeds the shaker cooling effect will be reduced, hence it is recommended that the engine should not operate for prolonged periods at reduced speeds.

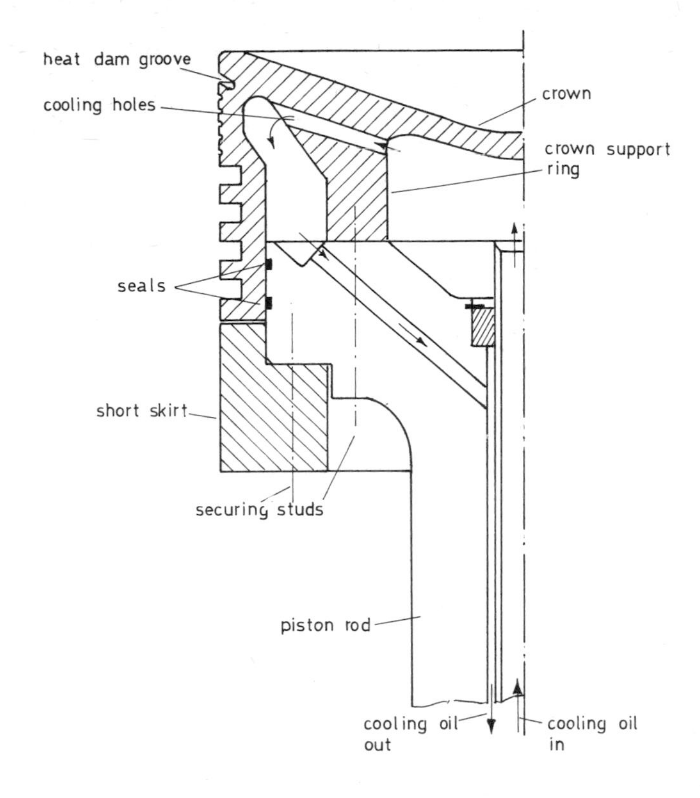

Fig. 34

For engines of moderate power rating oil cooling of pistons is normally used. Fig. 34 depicts in half section a oil cooled piston, note the way in which the oil is directed to flow radially across the underside of the crown, the method of supporting the crown and grooves to minimise thermal distortion.

Failure of pistons due to Thermal Loads

When a piston crown is subjected to high thermal load, the material at the gas side attempts to expand but is partly prevented from doing so by the cooler metal under and around it. This leads to compressive stresses in addition to the stresses imposed mechanically due to the variation in cylinder pressures.

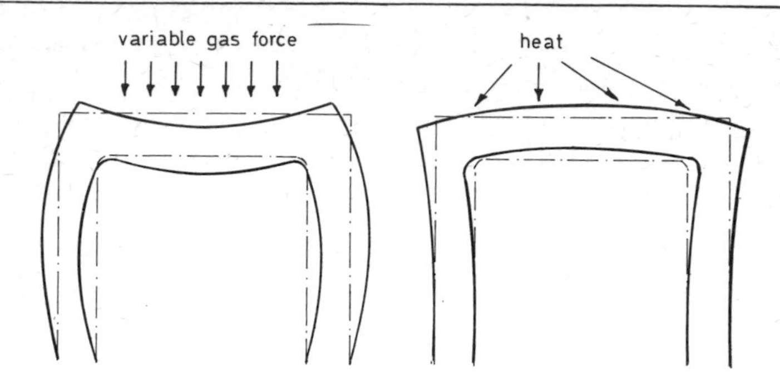

Fig. 35

At very high temperatures the metal can creep to relieve this compressive stress and when the piston cools a residual tensile stress is set up hence residual thermal stress. If this stress is sufficiently great, cracking of the piston crown may result.

At normal working temperatures the piston and cylinder liner surfaces should be parallel. Since there is a temperature gradient from the top to the bottom of the piston, allowance must be made during manufacture for the top cold clearance to be less than the bottom. The temperature gradient is generally non-linear and thermal distortions produce tensile stresses on the inner wall of the piston, gas forces tend to bulge the piston wall out thereby reducing the tensile stress. This variable tensile stress at very high thermal loads could lead to cracks propagating through from the inside of the piston to the piston ring grooves.

Piston Rings

Properties required of a piston ring:

1. Good mechanical strength, it must not break easily.
2. High resistance to wear and corrosion.
3. Self lubricating.
4. Great resistance to high temperatures.
5. Must at all times retain its tension to give a good gas seal.
6. Be compatible with cylinder liner material.

The above properties are the ideal and therefore difficult to achieve in practice. Materials that are used to obtain as many of the desired properties as possible are as follows:

1. Ordinary grey cast iron, in order that it may have good wear resistance and self lubricating property it must have a large amount of graphite in its structure. This however reduces its strength.

2. Alloyed cast iron, elements and combinations of elements that are alloyed with the iron to give finer grained structure and good graphite formation are: Molybdenum, Nickel and Copper or Vanadium and Copper.

3. Spheroidal Graphitic iron, very good wear resistance, not as self lubricating as the ordinary grey cast iron. These rings are usually given a protective coating, *e.g.* chromed or aluminised, etc. to improve running-in.

It is possible to improve the properties by treatment. In the case of the cast irons with suitable composition they can be heat treated by quenching, tempering or austempering. This gives strength and hardness without affecting the graphite.

Manufacture

1. Statically cast in sand moulds to produce either a drum from which a number of piston rings would be manufactured or an individual ring.

2. Centrifugally cast to produce a fine grained non-porous drum of cast iron from which a number of piston rings will be machined.

The statically cast rings, either drum or single casting, may be made out of round. The out of round blanks are machined in a special lathe that maintains the out of roundness. Rings manufactured in this way are expensive but ideal.

Most piston rings are made from circular cast blanks which are machined to a circular section on their inner and outer diameters. In order that the rings may exert radial pressure when fitted into the cylinders they are split in tension. Tensioning is done by cold deformation of the inner surface by hammering or rolling. The finished ring would be capable of exerting a radial pressure from 2 to 3 bar and have a Brinel hardness from 1600 to 2300 (S.I. units). Large diesel engine cylinder liners have a hardness range similar to the above.

Piston ring defects and their causes

1. Incorrectly fitted rings. If they are too tight in the grooves the rings could seize causing overheating, excessive wear, increased

blow-past, etc. If they are too slack in the grooves angular working about a circumferential axis could cause ring breakage and piston groove damage. If the butt clearance is too great, excessive blow-past will occur.

2. Fouling due to deposits on the ring sides and their inner diameters, this could lead to rings sticking, breakage, increased blow-past and scuffing.

3. Corrosion of the piston rings can occur due to attack from corrosive elements in the fuel ash deposits.

4. If the ring bearing surfaces are in poor condition or in any way damaged (this could occur during installation) scoring of cylinder liner may take place, if the ring has sharp edges it will inhibit the formation of a good oil film between the surfaces.

Due to uneven cylinder liner wear the piston ring diameter changes during each stroke, this leads to ring and groove wear on the horizontal surfaces. This effect obviously increases as differential cylinder liner wear increases. Oscillation of the piston rings takes place in the cycle about a circumferential axis approximately through the centre of the ring section, and if the inner edges are not chamfered they can dig into the piston groove lands. Keeping the vertical clearance to a working minimum will reduce the oscillatory effect.

In two stroke engines, piston rings have to pass ports in the cylinder wall. Each time they do, movement of segments of the rings into the ports can take place. This would be more pronounced if the piston ring butts are passing the ports. It is possible for the butts to catch the port edge and bend the ring. In order to avoid or minimise this possibility, piston rings may be pegged to prevent their rotation or they may be specially shaped.

Inspection of pistons, rings and cylinders

Withdrawal of pistons, their examination, overhaul or renewal. Together with the cleaning and gauging of the cylinder liner, is a regular feature of maintenance procedure. Frequency of which depends upon numerous factors, such as:

Piston size, material and method of cooling. Engine speed of rotation. Type of engine, 2- or 4-stroke. Fuel and type of cylinder lubricant used.

With high speed diesel engines of the 4-stroke type running time between piston overhauls is generally greater than that for large slow running two stroke engines. This can be attributed to the facts that: the engine is usually unidirectional, hence reduced numbers of stops and starts with their attendent wear and large fluctuations of thermal conditions. Small bore engines are easier to cool, cylinder volume is proportional to the square of the cylinder diameter hence increasing the diameter gives greatly increased cylinder content and high thermal capacity. Thus overhaul time can vary between about 2000 to 20 000h.

Pistons and cylinder liners on some engines can be inspected without having to remove the piston.

After scavenge spaces have been cleaned of inflammable oil sludge and carbon deposits, each piston can in turn be placed at its lowest position. The cylinder liner surfaces can then be examined with the aid of a light introduced into the cylinder through the scavenge ports. The cylinder liner surfaces should have a mirror-like finish. However, black dry areas at the top of the liner indicate blow past of combustion gases. Dull vertically striped areas indicate breakdown of oil film and hardened metal surface (this is caused by metal seizure on a micro scale leading to intense heating).

After inspection of a cylinder the piston can be raised in steps in order to examine both the piston and the rings. Heavy carbon deposits on piston crown and burning away of metal would indicate incorrect fuel burning and poor cooling. Piston rings should be free in the grooves, have a well oiled appearance, be unbroken and worn smooth and bright on the outer surface. If they are too worn then sharp burrs can form on the edges which enable them to act as scraper rings, preventing good oil film formation.

Poppet exhaust valves

The advent of residual fuel burning has led to a greater incidence of failure of exhaust valves, working surfaces have been damaged in various ways all of which result in gas blow past.

Causes of failure could be:—

1. Incorrect fuel burning leading to deposits on the working surfaces and burning of metal.

2. Ash deposits. The high ash content of residual fuel causes increased deposit on the valve working faces, this in addition to preventing the valve reseating can cause high temperature corrosion.

3. Incorrect valve timing.

4. Incorrect valve cooling.

Some of the methods that have and are being used to prolong exhaust valve life are:

1. Use of corrosion resistant hard metals for valve and seat. Generally in the form of a fused on insert.

2. False seats — not fused on — to avoid radial cracking due to thermal stresses.

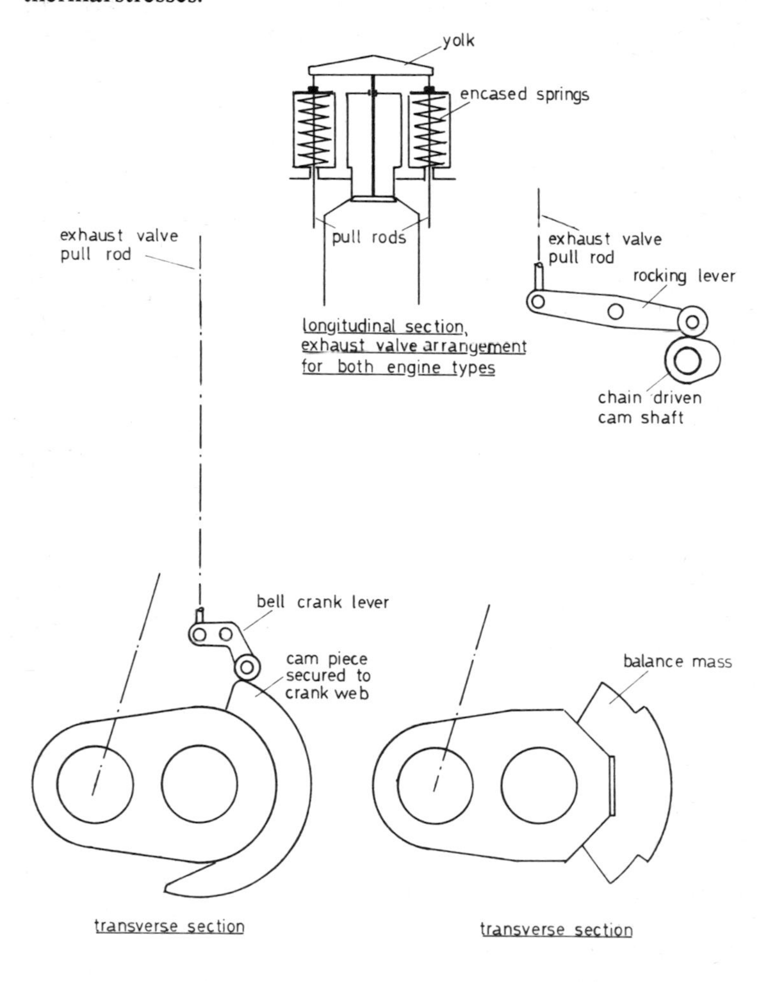

Fig. 36

3. Reducing valve temperature, this is possibly the most effective way of preventing valve burning. This can be done by (a) forced cooling around the exhaust valve, (b) increasing the scavenge period — this can result in a drop in exhaust valve temperature of about 70°C.

4. Rotating the valves on their seats by means of roto-caps. They employ the oscillating motion of the rocker arm converting it into intermittent rotary motion, this makes the settling of deposits between valve and seat difficult.

Various exhaust valve drives are used, the aim being to achieve reliability and good timing with efficiency. Most are mechanical but some are hydraulic.

Gotaverken Engine Exhaust Valve Drives. Fig. 36

The operation of the centrally arranged exhaust valve is controlled by cam segments fitted to the crank webs. Rollers, levers, and pull rods transmit movement to the yoke into which the exhaust valve spindle is fitted. Closing of the valve is accomplished by heavy encased springs fitted between the cylinder cover and the yoke.

For the more highly rated engine, the crankshaft diameter is larger, because of this and the fact that good accessibility to the main bearings would be affected the cams on the crankshaft were not retained. Accordingly, a separate camshaft is used for exhaust valve operation.

The exhaust valve is designed to give least possible resistance to gas flow in order to minimise energy loss. This energy saved would be used to effect in the turbo-charger.

TEST EXAMPLES

First class

1. Sketch and describe a main engine crankshaft for a diesel engine.

Explain how the wear down of the main bearings is checked by means of a deflection gauge. Give a set of typical values obtained from crankshaft deflections which would indicate that the crankshaft can be run safely without any undue stress.

2. Discuss the forces acting on the main engine crankshaft of a direct drive diesel when the engine is in service. Give possible reasons for (a) failure due to overstress, (b) failure without overstress.

3. Describe how you would conduct a survey of a large marine diesel engine crankshaft and bedplate. Discuss possible defects that may be found and causes.

4. Discuss the factors involved in the increase in the power/weight ratio in recent years of large marine diesel engines. What factors are presenting problems which limit further development.

5. Discuss the problems involved in the design of piston rings and explain the effects on the engine due to broken piston rings.

6. Discuss bearing clearances for (a) bottom end, (b) top end, (c) guides. What would be the effects on the engine if these clearances where not correctly maintained?

Second Class

1. Explain what would cause a relief valve to lift. Sketch such a valve and describe its overhaul.

2. Sketch and describe a piston for a large two stroke marine diesel engine, state the materials used in the construction and explain how the piston rings are fitted.

3. State the causes of crankshaft misalignment and describe how crankshaft deflections are taken.

4. Sketch and describe a main engine cylinder liner. Show the position of the water seals and describe how expansion is allowed for.

5. Sketch and describe a diesel engine internal lubrication arrangement. Why is the crosshead bearing difficult to lubricate? Give an example of an efficiently lubricated crosshead.

6. Discuss the causes of cylinder liner wear. Where would you expect the maximum wear to take place and why? Discuss the consumption of cylinder lubricating oil.

CHAPTER 3

FUEL INJECTION

DEFINITIONS AND PRINCIPLES

Atomisation

The break up of the fuel into minute spray particles so as to ensure an intimate mixing of air and fuel oil. This will allow maximum availability of oxygen contact with fuel droplets so giving complete combustion and maximum heat release from the fuel.

Turbulence

A swirl effect of air charge in the cylinder which in combination with atomised fuel spray gives intimate mixing and good overall combustion. Requires to be 'designed into' the engine by attention to liner, piston, ports, etc., details together with air pressure-temperature gradients.

Penetration

Ability of the fuel spray droplets to spread across the cylinder combustion space so as to allow maximum utilisation of volume for combustion.

Impingement

Excess velocity of fuel spray causing contact with metallic engine parts and resulting in flame burning.

Sprayer Nozzle

The arrangement at the fuel valve tip to direct fuel in the proper direction with the correct velocity. If the sprayer holes are too short the direction can be indefinite and if too long impingement can occur. If the hole diameters are too small fuel blockage (and impingement) can take place, alternatively too large diameters would not allow proper atomisation. In practice each manufacturer has a specific design taking into account method of injection, pressure, pumps, etc. Early Doxford practice utilised seven 0.6 mm holes. Sulzer eight 0.35 mm holes. B. & W. used four 1 mm holes, one hole directed up against air swirl and three holes directed down, with their opposed piston engine. As a generalisation the sprayer hole length: diameter ratio will be about 4:1, maximum pressure drop ratio about 12:1 and fuel velocity through the hole about 250 m/s.

Pilot Injection

A means of slowing down the rate of injection to prevent excessive cylinder pressure rise and 'diesel knock.' A small quantity of fuel as a pilot charge is introduced into the cylinder so that ignition occurs without excess pressure and the main charge following up is ignited readily without ignition delay.

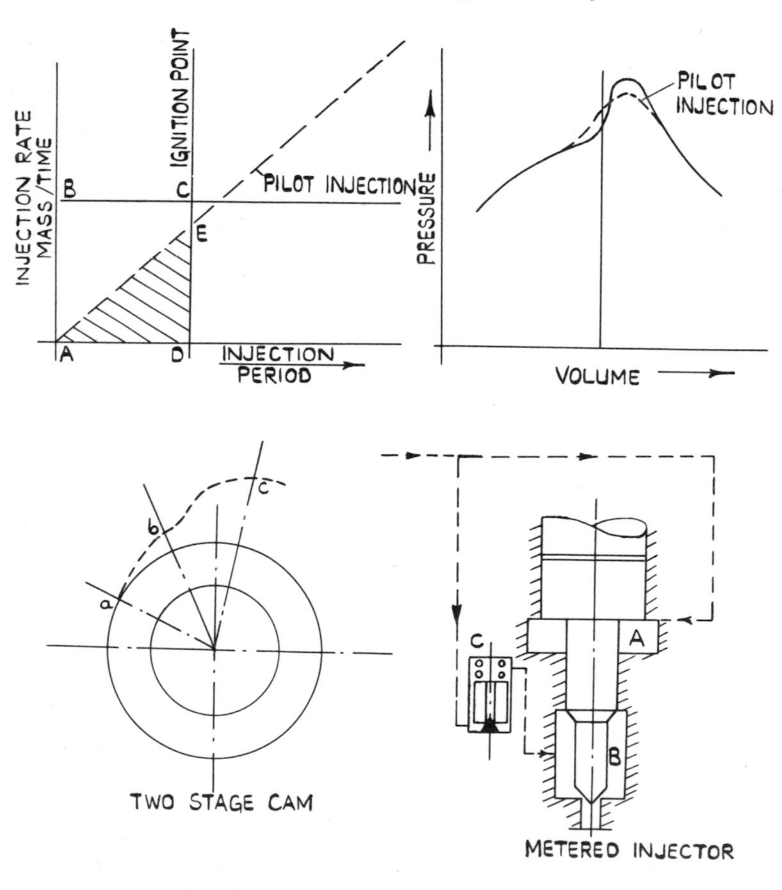

PILOT INJECTION DETAILS

Fig. 37

Referring to Fig. 37 the mass of fuel in the cylinder at injection point is proportional to area AED for pilot injection whereas for a quick opening fuel valve (for example hydraulic-spring injector) the corresponding area is ABCD. This illustrates the principle as does consideration of the draw card also shown in Fig. 37. A common system of achieving pilot injection used with high speed engines is also illustrated in Fig. 37. The two stage cam gives low pressure and low speed injection from a to b whilst both are progressively increased in the second stage from b to c. Fuel entering the cylinder between b and c burns immediately and the injection will cease at c. The cam is used in conjunction with the metered rise injector shown. Fuel pressure reaches annular space A and acts on the larger differential area so gradually opening the valve against the spring at a low pressure. Fuel pressure reaches the lower space B almost simultaneously as the non-return valve C is fairly lightly loaded. At the end of injection fuel pressure drops quickly in space A due to a spill action in the supply line. The non-return valve C also closes rapidly and the falling block piston displaces a big volume so dropping the pressure in the space B quickly. This means the needle has a metered gradual rise but a rapid snap shut action which prevents dribble. Valve lift cards of Fig. 12 (Chapter 1) illustrate pilot injection with Doxford engines and the mechanical fuel valve of Fig. 39 later in this chapter completes the discussion relating to this type of engine. An alternative method of achieving pilot injection with a metered dashpot type of hydraulic spring injector is also given later in this chapter.

Viscosity

May be defined as internal fluid molecular friction which causes a resistance to flow.

Pre-Heating

Refers in IC engine practice to the necessity of heating the oil at discharge pressure to ensure easy flow and good atomisation. This is generally necessary with boiler oils for use in IC engines as heating under suction conditions is limited by the flashpoint and appropriate DoT regulations apply. Degree of pre-heat depends on the initial viscosity of the oil and the viscosity required at the injector. Viscosity must be under 300 seconds (Redwood No.1) at injectors and preferably under 100 seconds. An oil of 1600 seconds (Redwood No. 1) requires heating to 100°C, 3500 seconds to 120°C and steam heating tracer pipes on fuel discharge would be necessary.

Boiler Oil

As previously mentioned these oils if burned in IC engines require engine modifications. Apart from improved purification, pre-heating, storage, circulation, etc., the fuel pump plunger clearances are increased, sprayer nozzle holes on fuel valves are usually (not always) of greater diameter and fuel quantities and valve lifts are greater (lower calorific value of boiler fuel requires a greater oil quantity to be burned).

Jerk Injection

Where the fuel pressure is built up at a fuel pump in a few degrees of rotation of the cam operating the plunger. Fuel is delivered directly to spring loaded injectors which are hydraulically opened when the jerk pump plunger lift has generated sufficient fuel pressure.

Common Rail

A system in which fuel pumps deliver to a pressure main and various cylinder valves open to the main and allow fuel injection to the appropriate cylinder. Requires either mechanically operated fuel valves (*e.g.* older Doxford engines) or mechanically operated timing valves (*e.g.* modern Doxford engines) allowing connection between rail and hydraulic injector at the correct injection timing.

Timed Injection

As defined above in which the fuel pump delivers to the timing valve and thence to the spring loaded injector. No lost motion clutch is required as the cam does not drive a pump plunger but operates a valve. The cam is symmetrical with respect to engine dead centre.

Note

Many aspects of fuels are covered in Volume 8 (Chapter 2) and revision of oil tests as well as basic definitions relating to specific gravity, Conradson carbon residue, Cetane number, etc., is strongly advised.

Indicator Diagrams

Details have been given of some typical indicator diagrams showing engine faults in Chapter 1. Aspects of fuel injection faults included late and early injection (draw card), fuel valve lift diagrams, etc., as well as related details such as compression cards. Two further typical faults are as illustrated in Fig. 38.

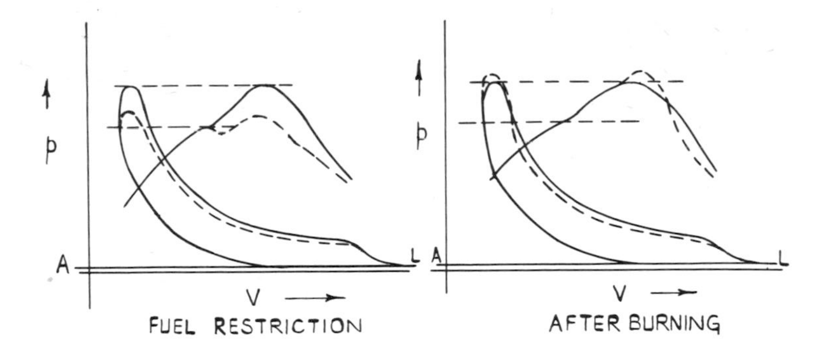

EFFECTS OF DEFECTIVE FUEL INJECTION

Fig. 38

Afterburning will show as indicated with a loss of power, increased cylinder exhaust temperature and possible discolouration of exhaust gases. Fuel restriction at filters, injectors, etc. or due to incorrect viscosity will result in a loss of power and reduced maximum pressure.

FUEL VALVES

Two types of fuel valve are still in use, *i.e.* mechanically operated and hydraulically operated. The latter have been used almost exclusively in recent years.

Mechanical Fuel Valve

Early designs were used on 4-stroke engines but this practice has now ceased. One type which may still be seen on some engines is the Doxford design. This is in fact adapted here for illustration because of some rather interesting principles of design and action. Refer to the diagrammatic sketch of Fig. 39.

The fuel valve lever fulcrums at the manoeuvring shaft lever, the fulcrum position being varied by rotation of the manoeuvring shaft by a direct mechanical linkage from the engine fuel control lever. As the roller moves to the right under the lifting action of the rotating cam a hardened pin moves the intermediate cam block to the left. The lifter lever (fulcrum about the sector shaft) causes the intermediate spindle to move left. Hydraulic pressure acting up through the pilot ram is allowed to pass by an external balance pipe to the fuel valve needle. The gap created by mechanical means at the intermediate spindle is filled by the hydraulic pressure moving the needle to the left on to the intermediate spindle face. Fine valve lift adjustments are also possible by either adjusting the screw nut at the fuel valve lifter or by varying the lifter fulcrum by adjustment of the eccentric sector shaft. (It is always

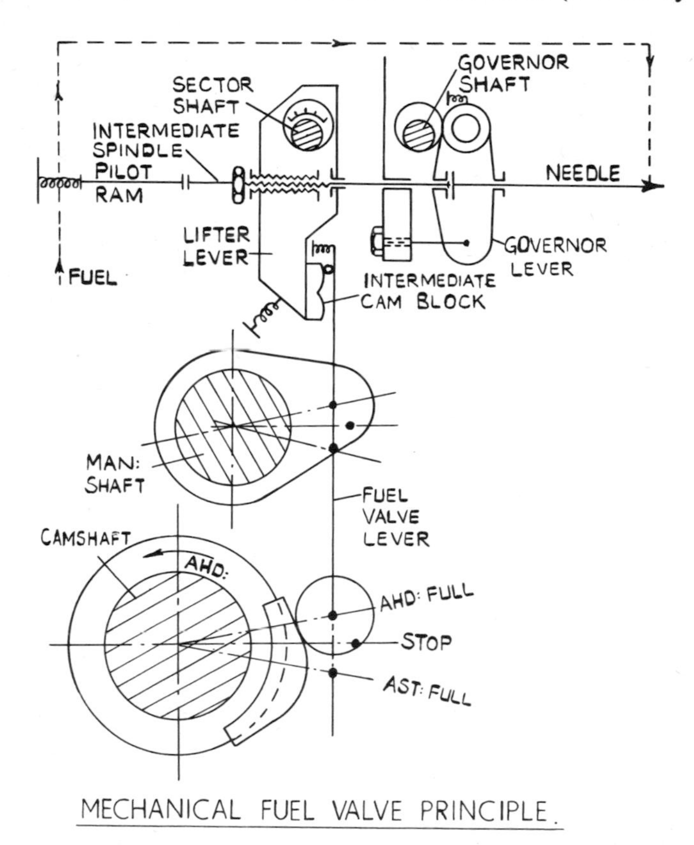

Fig 39

possible to increase lift by inserting liners under the main cam itself). The lift of the fuel valve is about 3 mm ahead (2 mm astern). The valve shown is the front valve. The back valve has an intermediate cam block with only one rising flank in place of two as shown. This means the striking pin does not touch the intermediate cam block in the astern position of the back manoeuvring shaft so the back valves do not operate when the engine runs astern. This valve, with a gradual lift action, gives pilot injection as discussed earlier in this chapter. A striker plate and pin to engage with the intermediate spindle allows fuel valve lift cards to be taken from the valve with a conventional indicator drive (typical diagram is shown in Chapter 1). When the engine governor (of the inertia type) operates, the action rotates the governor shaft so that the governor lever keeps the fuel valve needle from lifting. The governor also acts on the fuel pump to

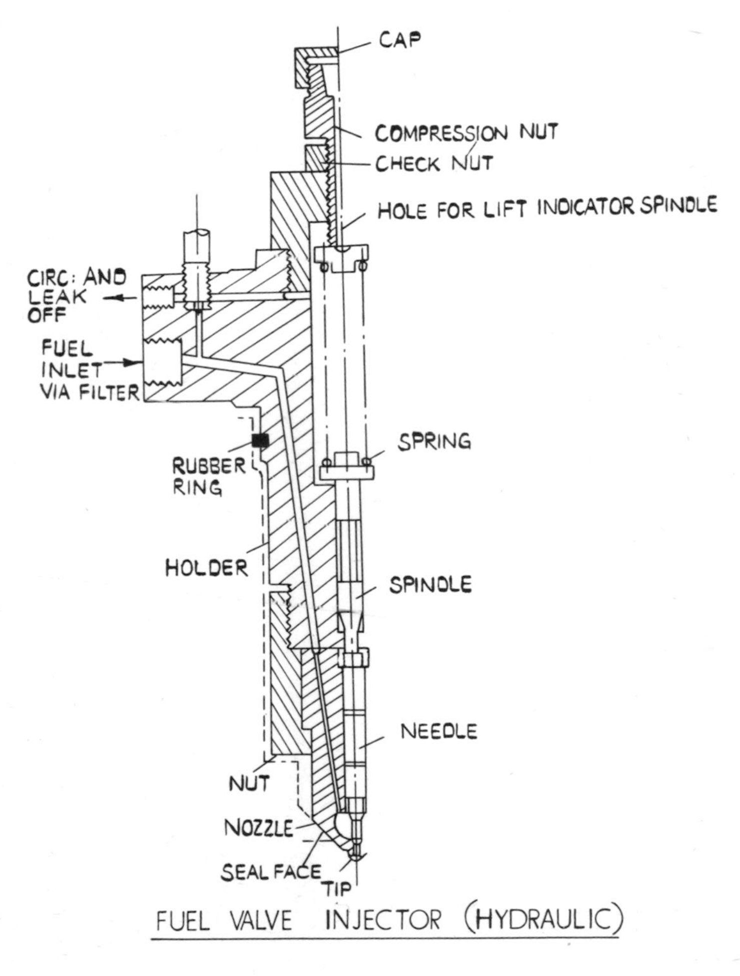

Fig. 40

cause spill on delivery with this engine. The needle itself is arranged in a housing (with water cooling) which bolts directly on to the cylinder liner, all the other mechanism shown is enclosed within a fuel valve cage which bolts on to the needle housing.

Hydraulic Fuel Valve

Fuel pressure acts on the needle in the lower chamber and when the force is sufficient to overcome the spring the needle lifts. Full lift occurs quickly as the extra area of the needle seat is exposed after initial lift. The full action of lift is limited by the needle shoulder which halts against a thrust face on the holder. The injector lift pressure varies with the design but may be about 140 bar average (some designs 250 bar).

Fuel valve lift diagrams for such an injector are given in Chapter 1. By removal of the spring cap the valve lift indicator needle can be assembled in the adaptor. This particular design as sketched is not cooled itself but is enclosed in an injector holder with seal (face to face) at tapered nozzle end with rubber ring at the top. Coolant is circulated in the annular space between the injector holder and the holder itself. Direct cooling of the fuel valve as an alternative to this is easily arranged. Coolant connections on the main block would supply and return through drillings similar to that shown for fuel. The choice of oil or water for cooling depends on the engine and valve design and is also affected by the type of fuel. With hot boiler oil it is probably necessary to utilise water and to arrange for cooling right to the injector tip so as to attempt to keep metal temperatures below 200°C. Coolant systems for fuel valves should be on a separate circuit. Hydraulic fuel valves usually have a lift of about 1 mm and the action is almost instantaneous. A metering effect with a controlled lift so as to approach a form of pilot injection can be achieved by using a controlled lift device with a dashpot. This is now described.

Controlled Lift Injector (Hydraulic)

A detail of this type of valve is given in Fig. 41.

Initial opening of the valve is controlled by a fairly light spring loading the needle in the conventional way. The top portion of the spring keep butts on to the spring casing which is held against the housing by the upper spring. This means the load of this spring does not act on the needle. An upper plunger is in contact with the spring casing and has a downward force exerted by fuel at discharge pressure on the plunger top. At slow engine speeds, with gas operated fuel pumps, the fuel discharge pressure is insufficient to compress the upper spring. As engine revolutions increase the

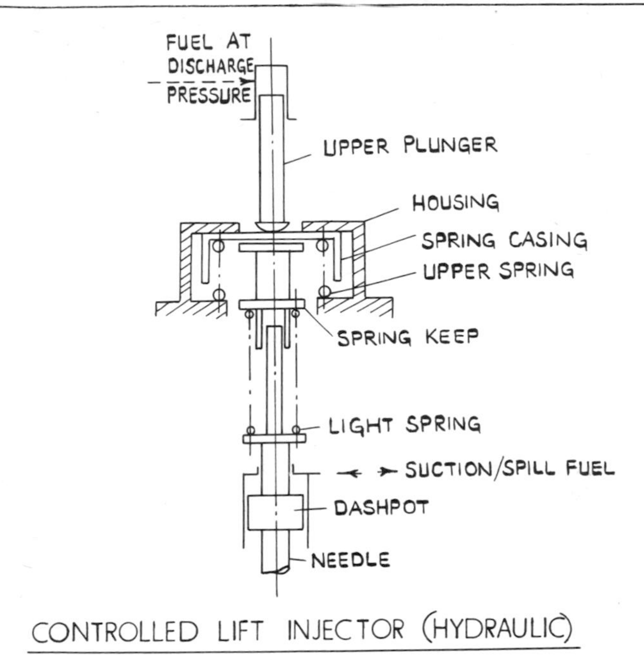

Fig. 41

pump delivery pressure will rise in direct ratio to the engine cylinder compression pressure. The plunger will force the spring casing down so increasing the light spring force and so raising injcction pressure. The opposite takes place with reduction in speed and leakage past the plunger is allowed for quick response in this case. Injection will commence at the correct engine piston position over the full range of engine revolutions from starting to full power. The dashpot is filled with fuel at atmospheric pressure and there is a small escape orifice at the top. The restriction and controlled rate of lift gives a small quantity of injection to start combustion with a progressively increased rate of injection up to the end of the delivery period. This is pilot injection, with a hydraulically operated fuel valve. The piston and needle are forced down after injection by the spring and oil is again drawn into the dashpot. The controlled rate of combustion gives a more even rise in cylinder pressure and less risk of Diesel knock.

The action of this valve is therefore twofold: (i) increased pressure with increased revolutions and load (consider the gas operated fuel pump given later in this chapter), and (ii) a pilot injection design with a controlled lift but quick return closure.

FUEL PUMPS

General

The physical energy demands of injection are great. Typical requirements include delivery of about 100 ml of fuel in 1/30 second at 750 bar so as to atomise over an area of 40 m^2. A peak energy input can reach 230 kW. A short injection period at high pressure, so placed to give the desired firing pressure, is necessary. Generally pilot injection and slow injection of charge is difficult to arrange for modern turbocharged engines.

Quantity Control

The amount of fuel injected per stroke can be varied by (1) mechanically varying the plunger stroke, (2) throttling the intake fuel, (3) varying the effective plunger stroke by cut-off or spill. The latter method is generally preferred and control is arranged to be either constant beginning on injection (delivery spill) or constant end of injection (variable start of delivery). Sometimes double controlled, which is a combination of the latter two techniques, has been used. Control has been arranged for regulated end of effective stroke by helical groove with constant beginning of injection, as in the well known Bosch principle, described later. This method is regularly utilised for auxiliary engines and gives fuel injection early in the cycle at light load which gives higher efficiency but also leads to higher firing pressures. It has also been utilised with large direct coupled engines (*e.g.* B. & W.). Control in valve type pumps for large engines was usually single controlled with constant end and regulation of the start of injection by varying the suction valve closure (*e.g.* Sulzer and Doxford). Engine performance at low load with later injection is a compromise between economy and firing pressure. Later Doxford designs utilised constant end but with helical suction control. With turbocharged engines the disadvantage of the constant end pump control is more noticeable as reduced firing pressure and efficiency is more marked at low loads due to reduced turbocharger delivery and pressure. In most designs the engine works satisfactorily but Sulzer have found it necessary to utilise double control. Tests with symmetrical or asymmetrical (bias to constant end or constant beginning) have been utilised. The latest RND type has double control in fact but the suction valve is not connected to the governor linkage and always closes at the same crank position so that effectively the simple constant beginning regulation has proved to be satisfactory.

Injection Characteristics

The diagram given in Fig. 42 illustrates some features of fuel injection based largely on Sulzer RND practice.

The fuel valve injector lift diagram is shown with lift of about 1.3 mm and injection period at full load approximately 6 degrees before to 22 degrees after. High firing pressures at full load in the high powered turbocharged range of engines can be reduced with a constant beginning method of injection. The ideal injection law corresponds to a rectangle with almost constant fuel pressure before the injector during injection. The practical curves shown show almost constant pressure at a reasonable maximum (750

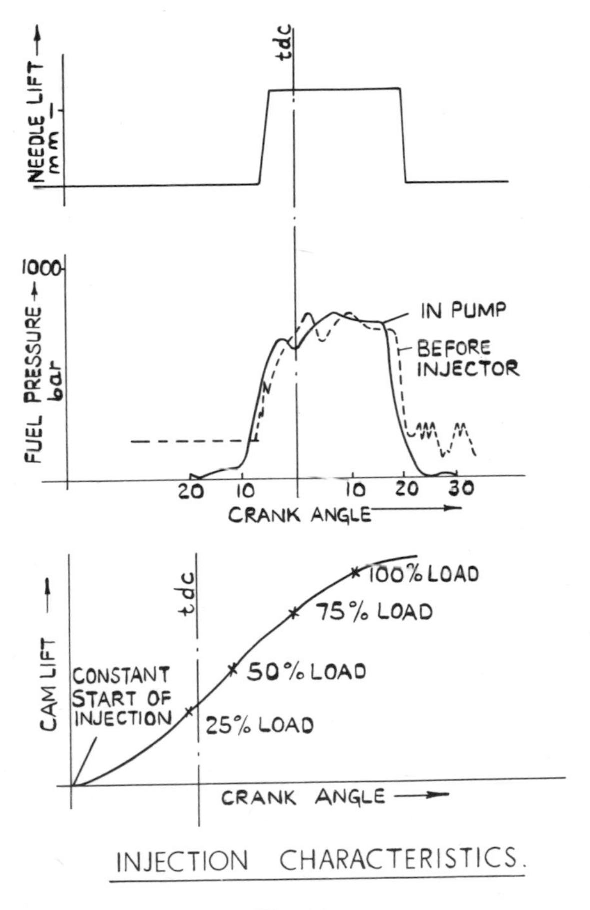

Fig. 42

bar). The last sketch of this Fig. 42 shows the effect of earlier spill at delivery for reducing load with a constant start of injection. The plunger at first moves very rapidly to build up pressure but is slowed during injection so giving minimum injection pressures at full load. The constant beginning of injection gives a flat combustion pressure over the full power range. At low loads the fuel pressure is higher than is usually the case because the plunger is delivering more in its maximum speed range, this gives better atomisation.

Bosch Jerk Pump Principle

The sketch numbered 1 on Fig. 43 shows the plunger rotated to be at about ¾ power position. On the upward delivery stroke fuel passes down the groove to the annular space beneath the helix.

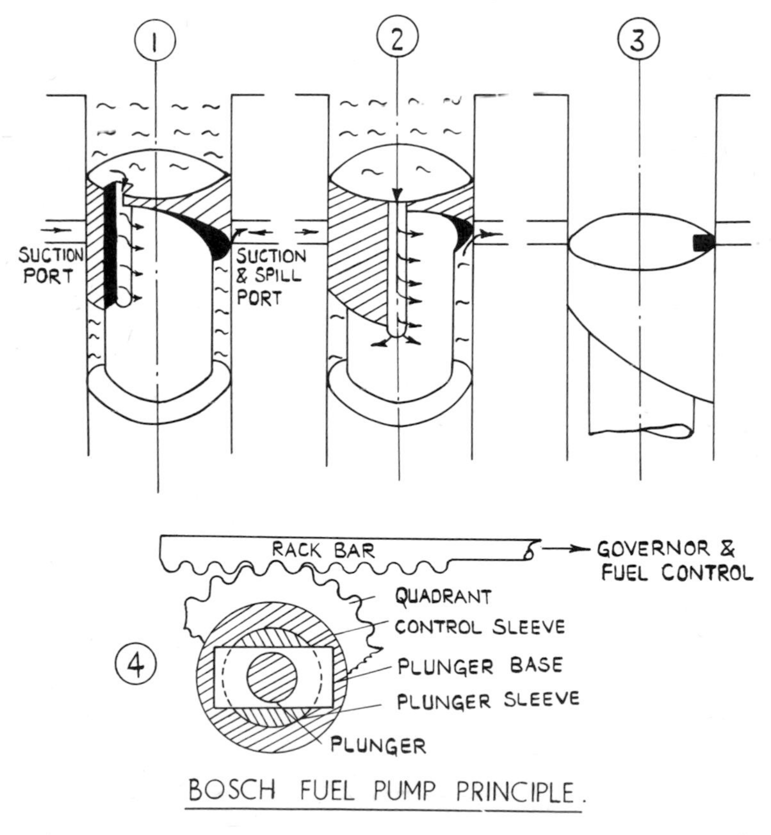

Fig. 43

Spill is about to occur as the helix uncovers the spill port. Sketch numbered 2 gives the position for about ½ power and rotation of the plunger to this position ensures an earlier spill due to the helix uncovering the spill port at an earlier point. The sketch numbered 3 illustrates zero load position with groove connected directly to spill. Sketch 4 shows a sectional plan view to illustrate how the plunger can be rotated so as to vary helix height relative to the spill port. The plunger base is slotted into a control sleeve which is rotated by quadrant and rack bar.

With both valve and helical spill designs there can be problems of fuel cavitation due to very high velocities. Velocities near 200 m/s can create low pressure vapour bubbles if pressure drops below the vapour pressure. These bubbles can subsequently

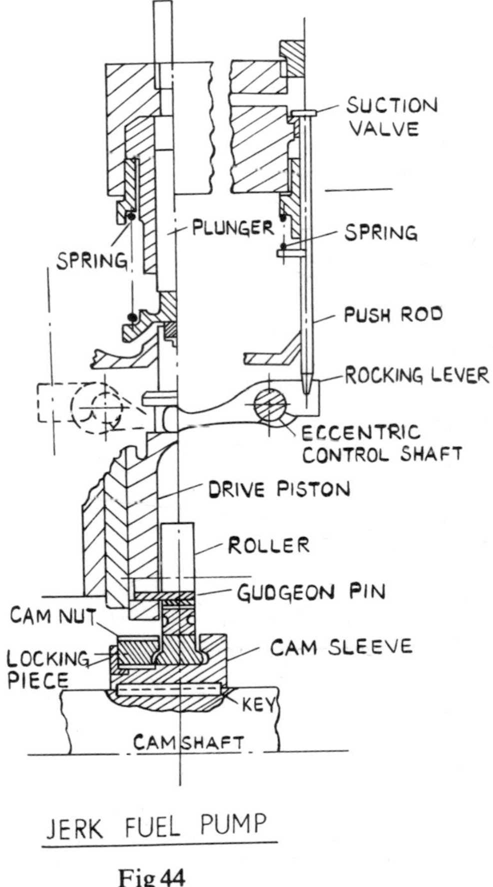

Fig 44

collapse during pressure changes which results in shock waves and erosion attack as well as possible fatigue failure. A spring loaded piston and orifice design can absorb and damp out fluctuations. Götaverken, Stork, MAN and B. & W. engines utilise a form of the above pump.

Jerk Fuel Pump Detail

The detail in Fig. 44 is based on Sulzer RD practice. Pumps are in pairs utilising one control shaft for both suction valves and with an upper block of forged steel and lower casing of cast iron. At the start of the upward delivery stroke the suction valve is open and effective stroke only commences when the suction valve closes. This means control on beginning of injection, which can be varied by the setting of the eccentric control shaft from fuel lever or governor, the end of injection is fixed when the roller mounts the cam dwell. The cam surface itself has a graduated width of track to equalise pressure between roller and cam to give non-slip operation. The parting gap between cam nuts and cams should measure about 0.2 mm on either side after tightening. The effective start of delivery for setting purposes is taken to be when the suction valve is 0.02 mm from closure. Shown dotted is a duplicate eccentric control shaft and rocking lever which is used with double regulation on later engine designs. This allows symmetrical double regulation (beginning and end) or asymmetrical with bias to one way. Plungers are often of nitrided steel and rollers of about 0.5% C steel, flame hardened.

Fuel Pump (Gas Operated)

Gas operated designs of fuel pump have been utilised for many years, certainly going back to conversion systems from blast injection. This Archaouloff system means a much reduced camshaft size with improved mechanical efficiency. By utilisation of gas operated lubricators a complete elimination of camshaft is possible. Referring to Fig. 45 it will be seen that gas enters through a regulated channel and has a double exit for quick clearance. Piston and plunger have an area differential of 10:1. Fuel spills at the beginning of injection until the suction valve closes. This point is variable by adjusting the fork lever control shaft from engine control or governor. Down movement of the piston under gas pressure first causes spill and then delivery commences until the plunger covers the discharge port (constant end). Oil trapped below the discharge port cushions the plunger and the return upwards by the spring expels gas. Actual injector timing is at the fuel valve and a suitable metered type of pilot injection valve has previously been described. In practice with a B and W type engine design injection timing was about 9 degrees before dead centre to 7 degrees after with this type of pump and associated fuel valve.

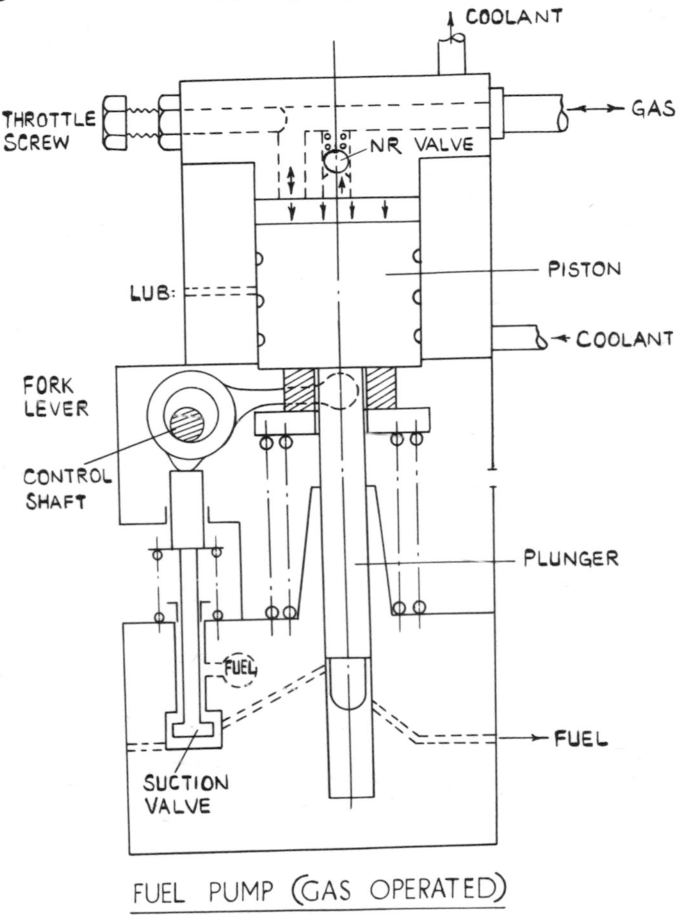

Fig. 45

FUEL SYSTEMS

There are obviously many and varied systems but the principles are in most cases the same. Two systems will now be described to cover firstly a type used with jerk pumps and secondly a type used with a timed injection system.

Oil Fuel System (Jerk Pumps)

The sketch shown in Fig. 46 is based on Sulzer RD practice but the principle is utilised with many other designs using jerk injection with spring loaded injectors.

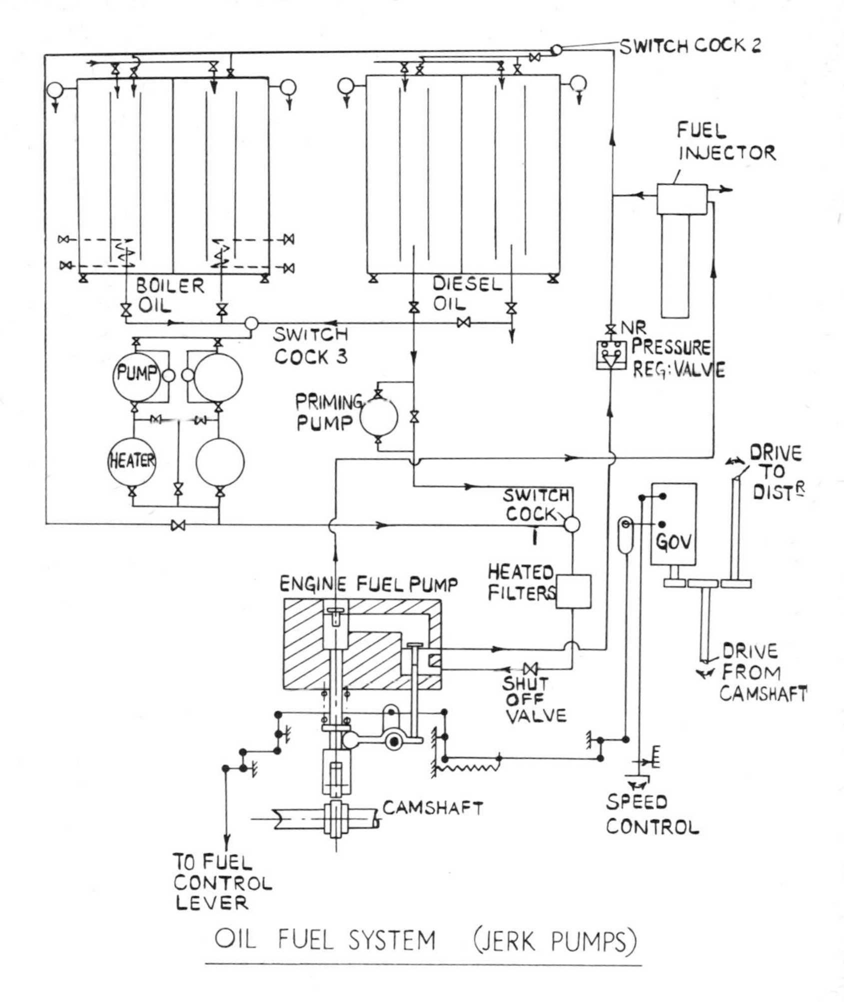

Fig. 46

The system is arranged to be suitable for manoeuvring with boiler oil and hence a good circulation and heating system are essential. Oil from the boiler oil service tank (provided with the usual filling, drain, run down, overflow, heating and vent arrangements) falls to the pump, with a relief byepass valve (setting at 13 bar) and thence to a switch cock (1) via an oil heater. The pump supplies excess oil to the engine fuel pump and the surplus passes back to the tank via a constant pressure valve maintaining suction side pressure at about 5 bar. Delivery from the fuel pump flows to the fuel injector and there is a priming spill and a recirculation back to the tank (sometimes with a cooler, not shown). Two other switch cocks (1 and 2) are fitted which allow diesel oil to the engine without oil circulation but with use of the priming pump. Heated fuel lines are necessary for boiler oil using steam tracer pipes. Engine fuel pump quantity control is varied from the engine controls, with adjustment of the beginning of injection, decided by the linkage to the suction valve shaft. The Woodward type governor is coupled by a slotted rod to the fuel pump linkage. Correct setting of the speed control handwheel, with appropriate clearances correctly adjusted, ensures direct governor control at all loads.

Oil Fuel System (Timed Injection)

Refer to Fig. 47.

Fuel is supplied to the suction manifold of the multi-plunger engine fuel pump at about 4 bar by a booster pump. Delivery at between 280 to 490 bar is delivered through a distribution block to the common rail. The engine fuel pump has diametrically opposed pump blocks, one per engine cylinder, with one eccentric throw driving the two opposed plungers. Control of fuel quantity is from the handwheel at the manoeuvring station, via linkages, control wheel shaft and spring link to the rack bar which varies the start of injection by a helical plunger device. A governor of the overspeed trip type is connected to the rack bar for direct action at the pump. Fuel delivery is taken from the distribution block to individual valve blocks with accumulator bottles, the latter providing extra volume in the system so damping out pressure variations. Injection of fuel, by the two spring loaded injectors per cylinder, is controlled by timing valves in pairs which are cam driven and control of the effective valve lift, and hence engine speed, is decided by the lever whose fulcrum shaft position is varied to the setting of the manoeuvring lever at the control station. A fuel spill valve (pneumatically loaded) maintains rail pressure as decided at the controls with spill to booster suction, duplicate stand-by valve. A pneumatically operated priming pump (priming up to 105 bar and pre-starting to 280 bar) and circulating pump are provided for manoeuvring on heavy oil but an electric priming pump only is satisfactory for diesel oil manoeuvring. The timing valve driving cam is in three parts, with facility for adjustment of timing, driven by a cam carrier fixed to the camshaft. The roller via the lever operates the tappet which opens and closes two valves (0.4 mm clearance between the two) whilst control operated from the engine controls has an independent adjustment for each unit. The pneumatic pumps work on almost identical principles to the gas fuel pump previously described using air pressures to a control cycling valve of about 3 bar.

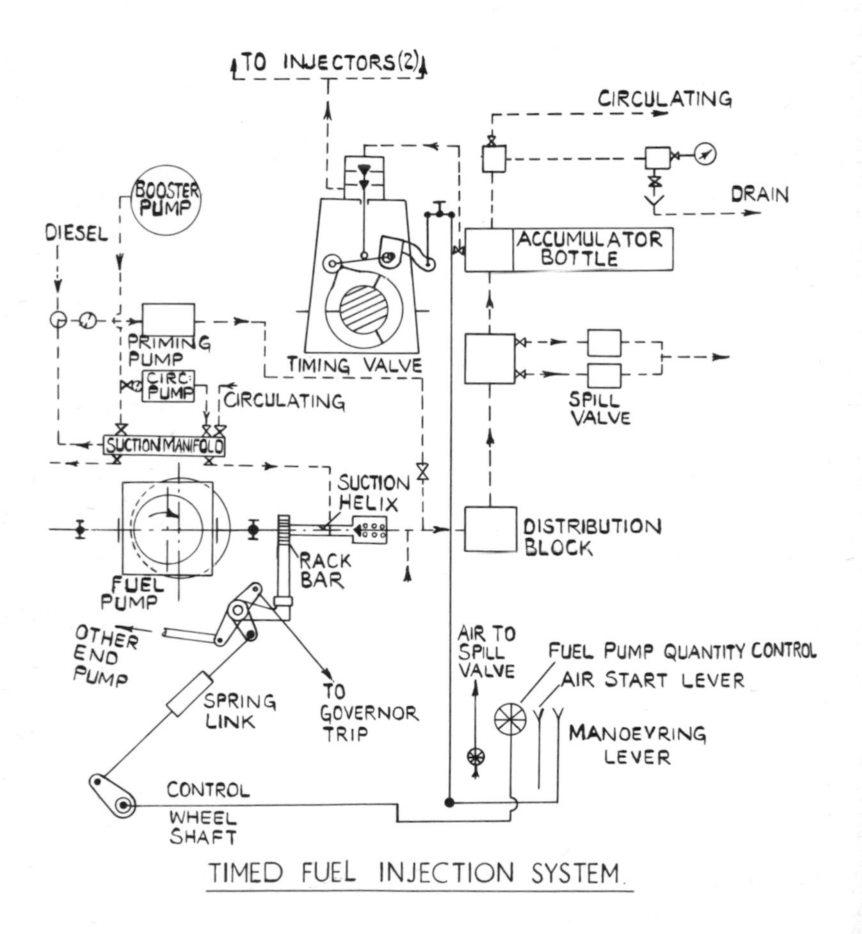

Fig. 47

TEST EXAMPLES

Second Class

1. With the aid of a sketch describe a Doxford fuel system which incorporates spring loaded fuel injectors. How is the fuel metered and how is the load in one cylinder balanced in this fuel system?

2. Discuss the fuel injection timing of a large IC oil engine under full and light load conditions. Describe how a reduction in the amount of fuel injected would be accomplished if the engine were equipped with:
(a) mechanically operated fuel valves,
(b) hydraulically operated fuel valves.

3. Sketch and describe a hydraulically operated fuel valve and describe how the fuel is metered into the engine.

4. With respect to compression ignition oil engine operation discuss the effects of:
(i) early firing
(ii) late firing
(iii) afterburning.
Show these effects on an indicator diagram sketch.

5. Discuss the following items with respect to the combustion of oil fuel in compression ignition oil engines:
(a) Atomisation, (b) Penetration, (c) Turbulence.
Consider carefully the design of a sprayer nozzle and comment on the proportions of spray hole diameter to length ratio.

6. With respect to fuel injection systems explain what is meant by:—
(a) common rail, (b) jerk injection.
Enumerate the possible advantages and disadvantages of each system. What is meant by the term 'pilot injection'?

First Class

1. Sketch and describe a mechanical fuel valve and explain how it is operated. Explain the working action. Discuss the arrangement adopted for astern running.

2. Sketch and describe a hydraulically operated fuel valve and explain how the opening and closing of the valve is timed. How could the pressure at which the fuel is injected be adjusted and are there any modifications which could be made to vary the opening rate? What attention do valves of this type require to keep them in good working order?

3. With respect to a jerk type of fuel pump:
(a) sketch the fuel cam profile and discuss the reasons for adaptation.
(b) illustrate with sketches how the fuel can be shut off by the overspeed governor trip action without alteration of the engine controls.

4. Sketch and describe a jerk type of fuel pump. With respect to such a pump show with explanatory sketches the initial method of timing the pump.

5. Draw indicator diagrams to illustrate the following:
(a) optimum full load conditions
(b) retarded fuel injection
(c) worn cylinder liner
(d) leaking fuel injector.

Discuss the effects of (b), (c) and (d) on the running of the engine.

6. Sketch and describe the Doxford timed fuel injection system. Explain how the fuel timing is arranged and quantity controlled.

CHAPTER 4

SCAVENGING AND SUPERCHARGING

If maximum performance and economy, etc. are to be maintained it is essential during the gas exchange process that the cylinder is completely purged of residual gases at completion of exhaust and a fresh charge of air is introduced into the cylinder for the following compression stroke. In the case of 4-stroke engines this is easily carried out by careful timing of inlet and exhaust valves, where because of the time required to fully open the valves from the closed position, and conversely to return to the closed position from fully open, it becomes necessary for opening and closing to begin before and after dead centre positions if maximum gas flow is to be ensured during exhaust and induction periods. Typical timing diagrams are shown in Fig. 48 for both normally aspirated and pressure charged 4-stroke engine types. Crank angle available for exhaust and induction with normally aspirated engines is seen to be of the order 420° to 450° with a valve overlap of 40° to 60° depending upon precise timing — with more modern pressure-charged engines this increases to around 140° of valve overlap. Basic object of overlap, *i.e.* exhaust and inlet valves open together, is to assist in final removal of remnant exhaust gases from cylinder so that contamination of charge air is minimal. The extension of overlap in the case of pressure-charged engines serves to: (1) further increase this scavenge effect, (2) provide a pronounced cooling effect which either reduces or maintains mean cycle temperature to within acceptable limits even though loading may be considerably increased. Consequent upon (2) it becomes clear that thermal stressing of engine parts is relieved and with exhaust gas turbo-charger operation prolonged running at excessively high temperatures is avoided. This latter would have an adverse effect on materials used in turbo-charger construction and could also contribute toward increased contamination.

In some cases an apparent anomaly exists between the temperature of exhaust gas leaving the cylinder and the temperature at inlet to turbo-charger, being as much as 90°C higher. This is partially explained by the fact that over the latter part of the gas

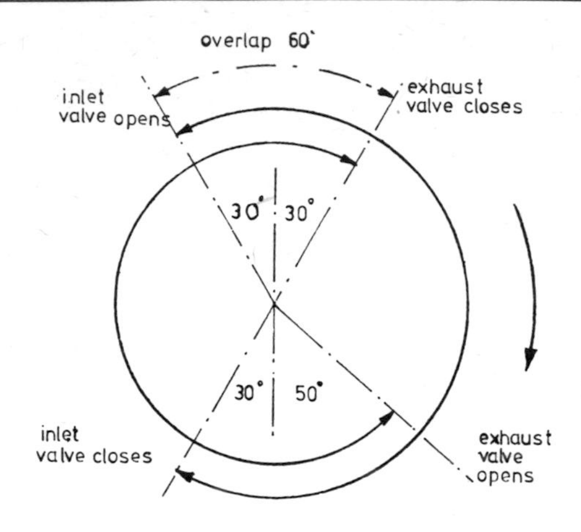

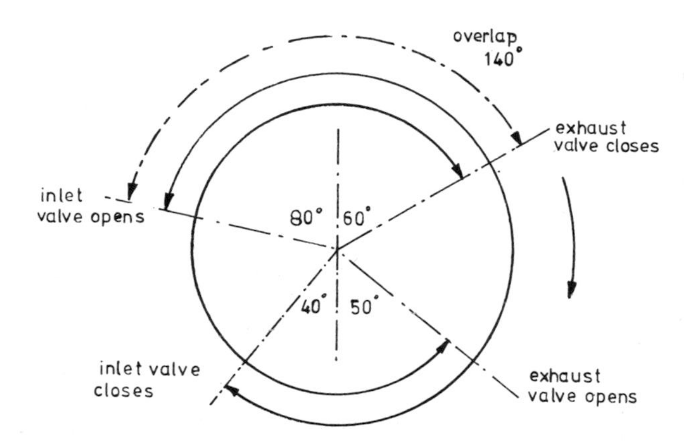

Fig. 48

exchange process the relatively cold scavenge air will have a depressing effect on the temperature indicated at cylinder outlet which will tend to indicate a mean value over the cyclic exchange. More probably the increase may be largely attributed to change of kinetic energy into heat energy and an approximately adiabatic compression of the gas column between cylinder and turbine inlet.

2-Stroke Cycle Engines

With only one revolution in which to complete the cycle the time available for clearing the cylinder of residual exhaust gases and recharging with a fresh air supply is very much reduced. Of necessity the gas exchange process is carried out around b.d.c. where the positive displacement effects of the piston cannot be exploited as is the case with the 4-stroke cycle. The total angular movement seldom exceeding 140° compared to well in excess of 400° with 4-stroke operation gives some indication of the need for high efficiency scavenging processes if cylinder charge is not to suffer progressive contamination and subsequent loss of performance with increased temperature and thermal loading. Prior to the introduction of turbo-charging to 2-stroke machinery this necessitated a low degree of pressure-charging of 1.1 to 1.2 bar to ensure adequacy of the gas exchange process.

The scavenging of 2-stroke engines is generally classified as: (1) uniflow or longitudinal scavenge, (2) loop and cross scavenge. Because of the simplicity of the arrangement uniflow scavenge as employed in poppet-valve or opposed piston type engines is generally considered as being the most efficient. In this case charge air is admitted through ports at the lower end of the cylinder and as it sweeps upwards toward the exhaust discharge areas, almost complete evacuation of residual gases is obtained. By suitable design of the scavenge ports or the provision of special air deflectors the incoming charge air can be given a swirling motion which intensifies the purging effect and also promotes the degree of turbulence within the charge which is required for good combustion when fuel injection takes place.

Cross and loop scavenge have both exhaust and scavenge ports arranged around the periphery of the lower end of the liner and in so doing eliminate the need for cylinder head exhaust valves, etc. and their attendant operating gear. This considerably simplifies engine construction and can lead to a reduction in maintenance. Because of simplified cylinder head construction the cylinder combustion space can be designed for optimum combustion conditions. Generally however, the scavenging efficiency is somewhat lower than with the uniflow system due to the more complex gas-air interchange and the possibility of charge air passing straight to exhaust with little or no scavenging effect. Careful attention to port design does however considerably reduce this problem.

The gas exchange process itself may be divided into three separate phases: (1) blowdown, (2) scavenge and (3) post-

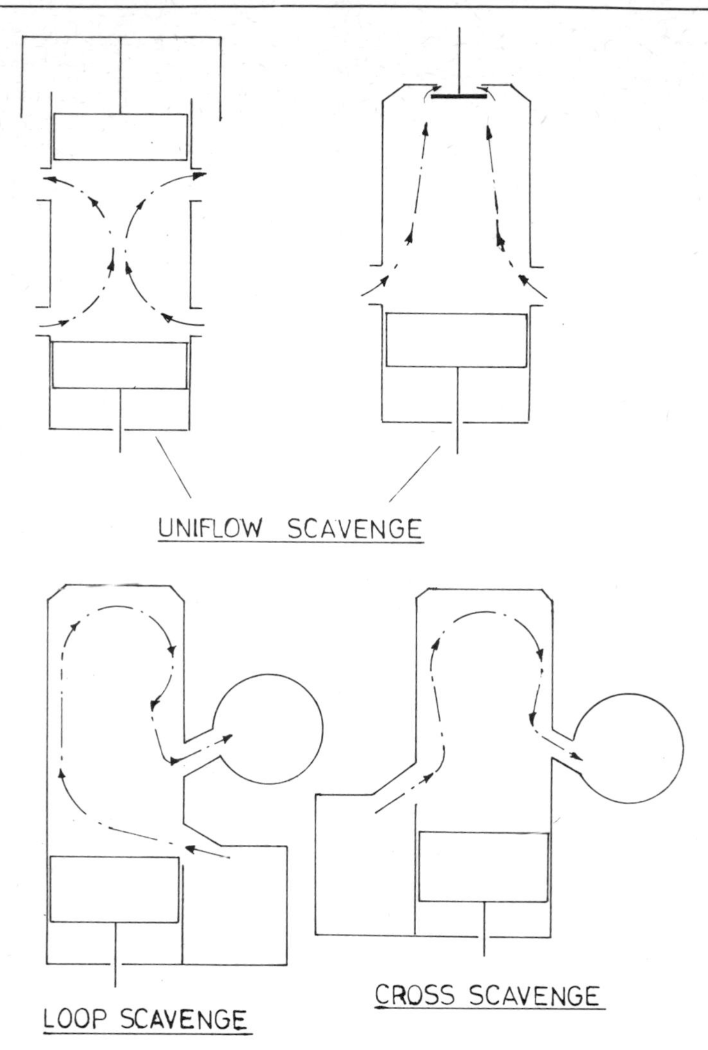

Fig. 49

scavenge. During blowdown the exhaust gases are expelled rapidly — the process being assisted by amply dimensioned ports or valves arranged to open rapidly. At the end of this blowdown period when the scavenge ports begin to uncover, the cylinder pressure should be at or below charge air pressure so that the scavenge process which follows, effectively sweeps out the remaining residual gases. With scavenge ports closed the post scavenge period completing the gas exchange process should ensure that exhaust discharge areas close as quickly as possible to prevent undue loss of charge air so that the trapped air at beginning of compression has the highest possible density. Although some loss of charge air is unavoidable it should be borne in mind that the air supply is considerably in excess of that required for combustion and the cooling effect of the air passing through the system has the

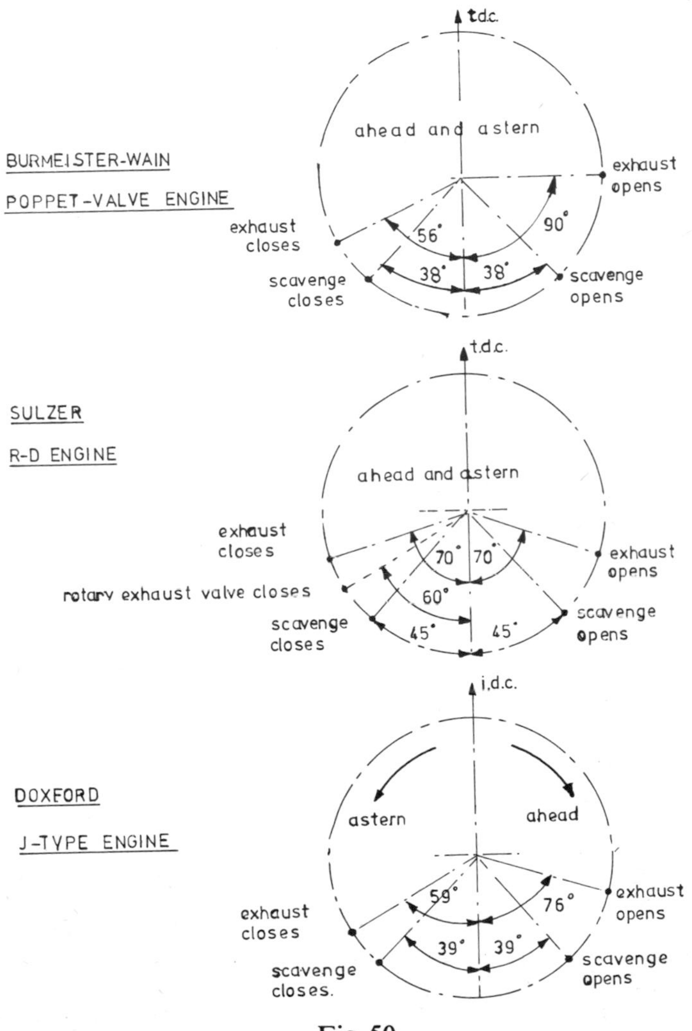

Fig. 50

result of keeping mean cycle temperatures down so that service conditions are less exacting. The increased cylinder pressures encountered with modern turbo-charged machinery may result in exhaust opening being advanced so that sufficient time is given for cylinder pressure to fall to or below charge air pressure when the scavenge ports uncover. A complementary aspect of earlier opening to exhaust is the increased pulse energy obtainable from

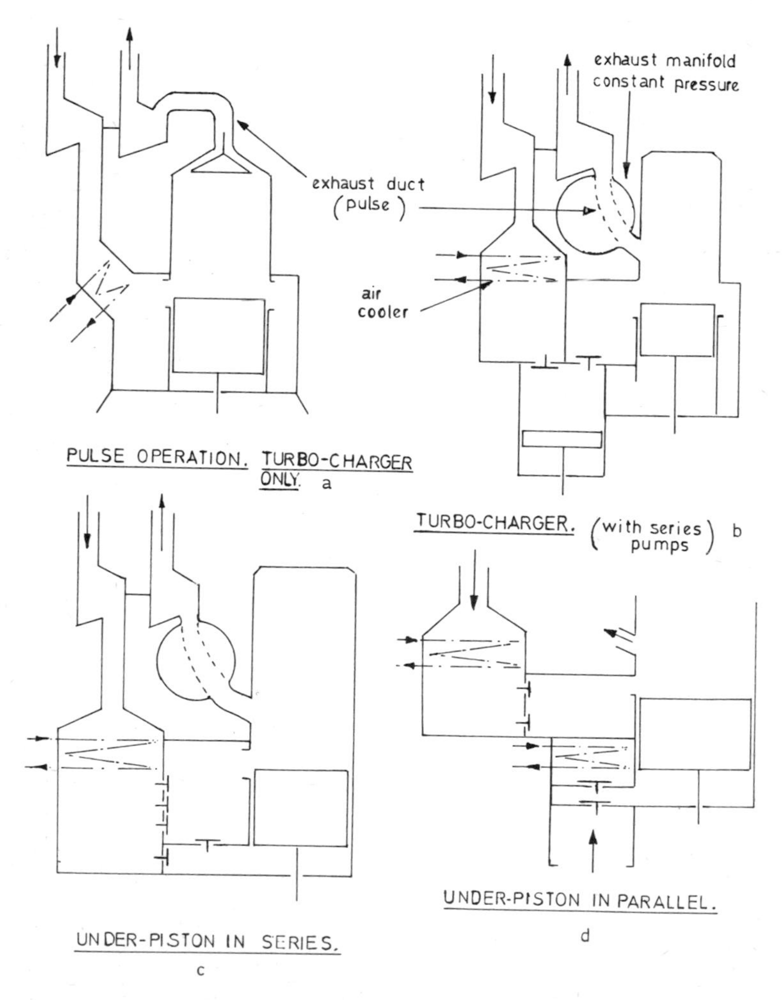

Fig. 51

the exhaust gas which can be utilised to improve turbo-charger performance. In many cases this is the main criterion which influences exhaust opening, since the loss of expansive working is more than offset by the gain in turbo-charger output. Obviously in the case of reversing engines there may be some slight penalty incurred if prolonged operation in the astern direction is considered. Fig. 50 shows the timing for some of the present generation of direct drive slow speed diesels.

PRESSURE CHARGING

By increasing the density of the air charge in the cylinder at the beginning of compression a correspondingly greater mass of fuel can be burned giving a substantial increase in power developed. The degree of pressure charging required, which determines the increase in air density, is achieved by the use of free running turbo-chargers which are driven by the exhaust gases expelled from the main engine. About 20% of the energy available in the exhaust gas is utilised in this way. It is also the usual practice to employ some form of scavenge assistance either in series or in parallel with the turbo-chargers. This may take the form of engine driven reciprocating scavenge pumps, under piston effect, independently driven auxiliary blower, etc.

Fig. 51 (b) and (c). Turbo-charger provides charge air at 70 to 95% of required pressure with under-piston effect or series pump making up the balance. Slight increase in temperature of air delivered to engine since air cooling is carried out after the turbo-charger only.

Fig. 51 (d). With parallel operation air supply to engine is increased by air delivery from pumps with proportionate increase in output resulting in greater exhaust gas supply to turbo-charger and improved turbo-charger performance.

The advantages of pressure-charging may be summed up as:(1) substantial increase in power for a given speed and size; (2) better mass power ratio, *i.e.* reduced engine mass for given output; (3) improved mechanical efficiency with reduction in specific fuel consumption; (4) reduction in cost per unit of power developed; (5) the increase in air supply has a considerable cooling effect leading to less exacting working conditions and improved reliability. Because of increasing power output diesel plant has now moved into the field which for years was dominated by the steam turbine.

Constant Pressure and Pulse Operation

In general the manner in which the energy of the exhaust gases is utilised to drive the turbo-charger may be ascribed to (1) the pulse system of operation and (2) constant pressure operation.

Pulse Operation

This makes full use of the higher pressures and temperatures of the exhaust gas during the blow-down period and with rapidly opening exhaust valves or ports the gases leave the cylinder at high velocity as pressure energy is converted into kinetic energy to create a pressure wave or pulse in the exhaust lead to the turbo-charger. For pulse operation it is essential that exhaust leads from cylinder to turbine entry are short and direct without unnecessary bends so that volume is kept to a minimum. This ensures optimum use of available pulse energy and avoids the substantial losses that could otherwise occur with a corresponding reduction in turbo-charger performance. Of necessity, exhaust ducting must be arranged so that the gas-exchange processes of cylinders serving the same turbo-charger do not interfere with each other to cause pressure disturbances that would affect purging and recharging with an adverse effect upon engine performance. With 2-stroke engines the optimum arrangement is three cylinder grouping with 120° phasing which gives up to 10% better utilisation of available energy than cylinder groupings other than multiples of three. Due to the small volume of the exhaust ducting and direct leading of exhaust to turbine inlet the pulse system is highly responsive to changing engine conditions giving good performance at all speeds. Theoretically turbo-charging on the pulse system does not require any form of scavenge assistance at low speeds or when starting. In practice however the use of an auxiliary blower or some other means of assistance is employed to ensure optimum conditions and good acceleration from rest.

Constant Pressure Operation

In this system the exhaust gases are discharged from the engine into a common manifold or receiver where the pulse energy is largely dissipated. Although the pulse energy is lost, the gas supply to the turbine is at almost constant pressure so that optimum design conditions prevail since, under normal conditions, gas flow will be steady rather than intermittent. Further, as engine ratings increase, the constant pressure energy contained in the exhaust gas becomes increasingly dominant so that sacrifice of pulse energy in a large volume receiver is of less consequence. Fig. 52 shows the results of tests carried out on a Sulzer type engine which indicates that up to b.m.e.p. of around 7 bar the advantage lies with the pulse system but as b.m.e.p.

increases beyond this figure the constant pressure system becomes more efficient giving greater air throughout and some slight reduction in the fuel rate.

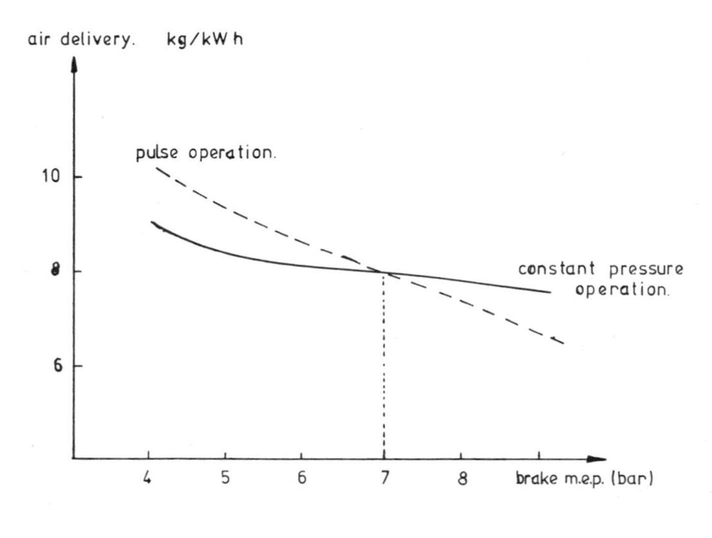

Fig. 52

Due to the much larger volume of the exhaust system associated with constant pressure operation the release of exhaust gas is rapid and earlier opening to exhaust is generally only necessary to ensure cylinder pressure has fallen to or below the charge air pressure when the scavenge ports begin to uncover. With a possible reduction in exhaust lead expansive working can be increased which is a further contributory factor in reducing the fuel rate. Major drawback to constant pressure operation is that the large capacity of the exhaust system gives poor response at the turbo-charger to changing engine conditions with the energy supply at slow speeds being insufficient to maintain turbo-charger performance at a level consistent with efficient engine operation. Some form of scavenge assistance is therefore essential. To offset this however the number of turbo-chargers required as compared to pulse operation can be reduced, a greater flexibility exists in the case of turbo-charger location and exhaust arrangement and no de-rating of engine need be considered for cylinder groupings other than multiples of three.

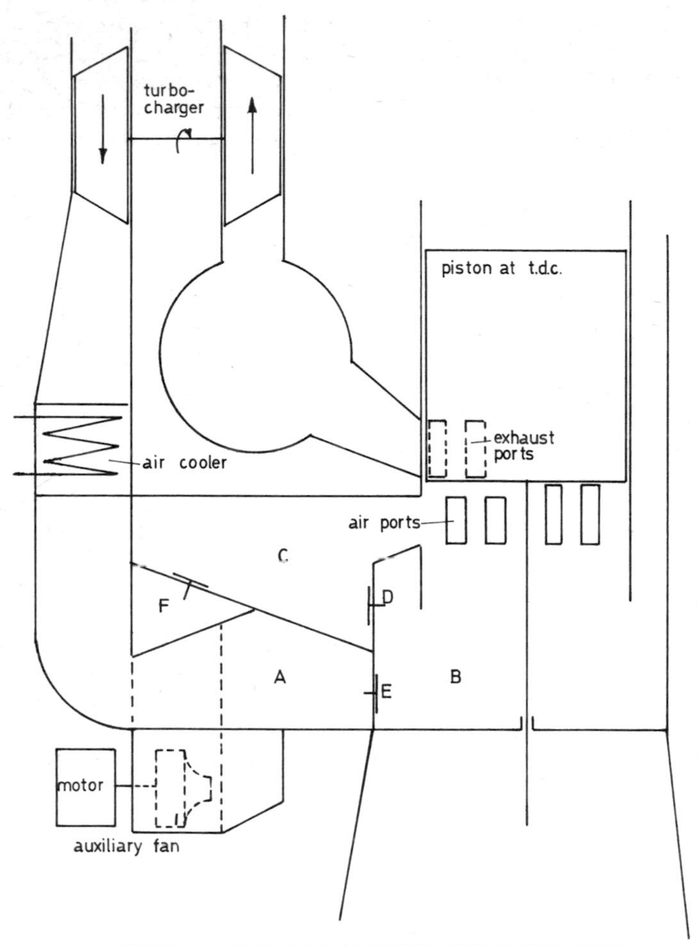

Fig. 53

Fig. 53 shows the diagrammatic arrangement of the Sulzer RND engine which operates with constant pressure supercharge. In normal operation air is drawn into under-piston space B from common receiver A and compressed on downstroke of piston to be delivered into space C so that when scavenge ports uncover purging is initiated with a strong pressure pulse. As soon as pressure in spaces B/C falls to common receiver pressure in A scavenge continues at normal charge air pressure. For part load operation the auxiliary fan is arranged to cut in when charging pressure falls below a pre-set value. Air is drawn from space A and delivered into space F and this together with under piston effect ensures good combustion and tιouble-free operation under transient conditions.

AIR COOLING

During compression of the air at the turbo-blower, which is fundamentally adiabatic, the temperature may increase by some 40°C with a corresponding reduction in density. This means that the air must be passed through a cooler on its passage to the engine in order to reduce its temperature and restore the density of the charge air to optimum conditions. Correct functioning of the cooler is therefore extremely important in relation to efficient engine operation. Any fouling which occurs will reduce heat transfer from air to cooling medium and it is estimated the 1°C rise in temperature of air delivered to the engine will increase exhaust temperature by 2°C. Reduction in air pressure at cooler outlet due to increased resistance is also a direct result of fouling. Under conditions of high humidity or where the moisture content of the ambient air is high, precipitation at the cooler may be copious so that adequate drain facility must be provided. In extreme circumstances it may be necessary to employ some alternative means of water separation if carry-over of moisture to the engine is to be avoided.

TURBO-CHARGERS

These are essentially a single stage axial flow turbine driving a single stage centrifugal air compressor via a common rotor shaft to form a self-contained free running unit. Expansion of the exhaust gas through the nozzles results in a high velocity gas stream entering the moving blade assemly. Because of the high rotational speeds perfect dynamic balance is essential if troublesome vibrations are to be avoided. Even with this, the effect of external vibrations being transmitted via the ship's structure to the turbo-charger is a further problem to be resolved. This is done by mounting the bearings in resilient housings incorporating laminar spring assemblies to give both axial and radial damping effect. Another aspect of this arrangement is to prevent flutter or chatter at bearing surfaces when stopped so that incidental bearing damage is prevented. Lubrication of the bearings may be by separate or integral oil feed, but whatever arrangement is adopted it must be fully effective at a steady axial tilt of 15° and support a temporary tilt of 22½° as may occur in a heavy seaway. The bearings themselves may be a combination of ball and roller bearings or separate sleeve (journal) type bearings.

The resistance to motion by plain bearings both when starting from rest and in normal operation, is much greater than that

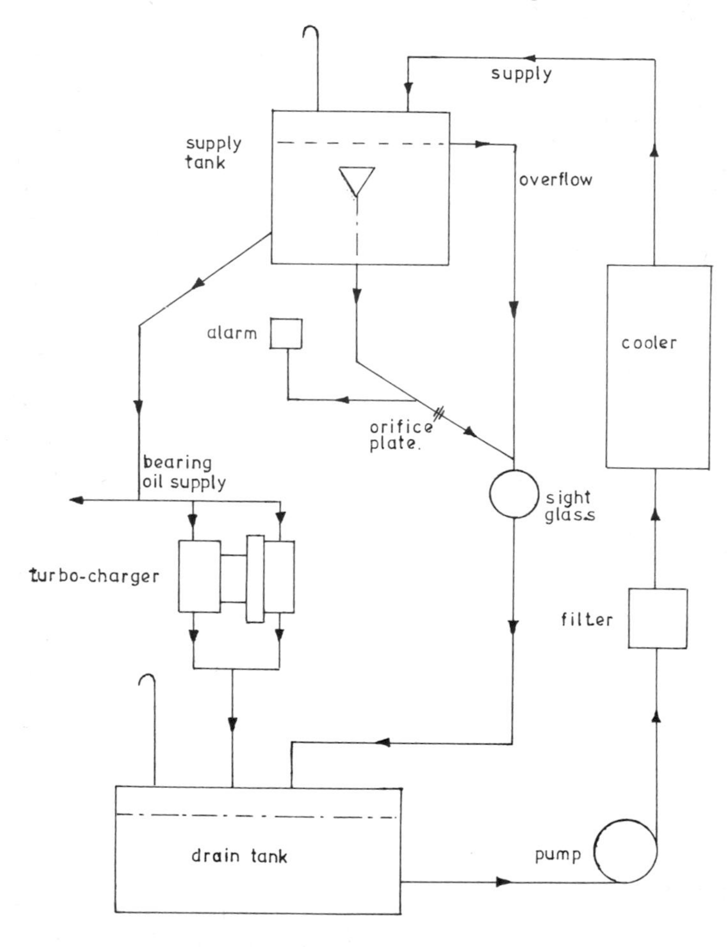

TURBO-CHARGER LUBRICATION SYSTEM

Fig. 54

encountered with ball and roller bearing assemblies. With high speeds of operation this is obviously a factor which favours the use of the latter type. Against this however is the fact that periodic replacement of ball and roller assemblies is essential if trouble-free service is to maintained — this is due to the fact that rapid and repeated deformation with resultant stressing causes surface metal fatigue of contact surfaces with the result that failure will occur. The effects of vibration, overloading, corrosion or possible abrasive wear, etc, lead to premature failure which emphasises the need for isolation of bearings from external vibrations together with use of correct grade of lubricant and effective filtration, etc. Plain bearings should however have a life equal to that of the blower provided that normal operating conditions are not exceeded.

For a separate oil feed as shown in Fig. 54 the oil level in the high level tank should be maintained about 6 m above the turbo-chargers. This will ensure that the oil pressure reaching the bearings should never fall below a pressure of around 1.6 bar. If level of oil falls below the mouth of the inner drain pipe it is quickly emptied and an alarm condition initiated. After an alarm it takes about ten minutes to empty the high level tank which is sufficient to ensure adequate lubrication of the turbo-chargers as they run down after the engine is stopped.

Referring to Fig. 55 it can be seen that the blower end of the turbo-charger consists of a volute casing of light aluminium alloy construction which houses the inducer, impeller and diffuser which are also of light alloy construction. The function of the inducer is to guide the air smoothly into the eye of the impeller where it is collected and flung radially outward at ever increasing velocity due to the centrifugal effect at high rotational speed. At discharge from the impeller it passes to the diffuser where its velocity is reduced in the divergent passages thus converting its kinetic energy into pressure energy. The diffuser also functions to direct air smoothly into the volute casing which continues the deceleration process with further increase in air pressure. From here the air passes to the charge air receiver via the air cooler.

The turbine end of the turbo-charger consists of water-cooled cast-iron exhaust gas casings which house the nozzle-ring turbine wheel and blading, etc. These together with the rotor shaft are of heat resisting alloy steel (nickel-chrome alloy) to withstand continuous operation at temperatures in excess of 450°C. Some degree of air cooling is given by controlled leak-off past the labyrinth seal between back of impeller and volute casing. An air

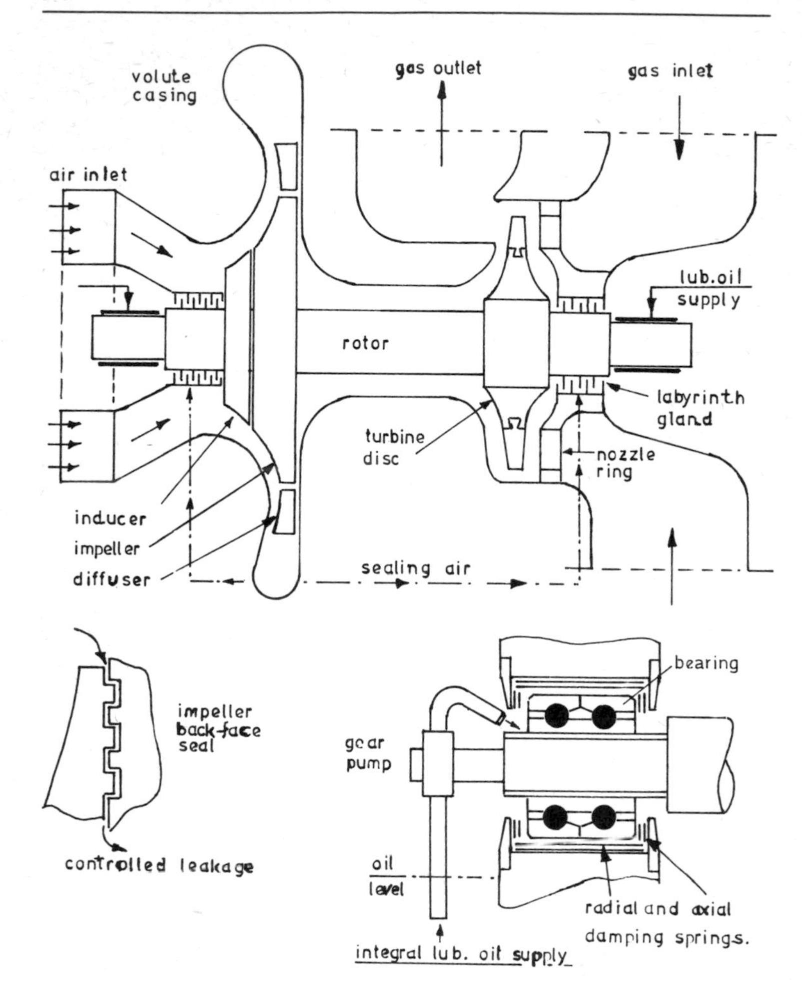

TURBO-CHARGER

Fig. 55

seal is also provided at the labyrinth glands next to the rotor bearings by air-bleed from the delivery side of the blower to prevent air/gas and oil tracking across the bearings.

Cooling media for the exhaust gas casings is generally from the engine jacket water cooling system although in some cases sea-water has been employed. In both cases anti-corrosion plugs are fitted to prevent or inhibit corrosion on the water side. With water cooled casings experience has shown that under light load conditions when low exhaust temperatures are encountered it is possible that precipitation of corrosive forming products — mainly sulphuric — will occur on the gas side of the casing. This results in serious corrosive attack which is more marked at the outlet casing because of lower temperatures. Methods of prevention such as enamelling and plastic coatings, etc. have been tried to alleviate this problem with varying degrees of success. A particularly effective approach to the problem is the use of air as the cooling media with the result that this particular instance of corrosive attack is virtually eliminated.

Water Washing — Blower Side

On the air side, dry or oily dust mixed with soot and a possibility of salt ingestion from salt laden atmosphere can lead to deposits which are relatively easy to remove with a water jet, usually injected at full load with the engine warm. A fixed quantity of liquid (1 to 2½ litre depending upon blower size) is injected for a period of from 4 to 10 seconds after which an improvement should be noted. If unsuccessful the treatment can be repeated but a minimum of ten minutes should be allowed between wash procedures. Since a layer of a few tenths of a millimeter on impeller and diffuser surfaces can seriously affect blower efficiency the importance of regular water washing becomes obvious. It is essential that the water used for wash purposes comes from a container of fixed capacity — under no circumstances should a connection be made to the fresh water system because of the possibility of uncontrolled amounts of water passing through to the engine.

Water Wash — Turbine Side

This is generally carried out at reduced speed by rigging a portable connection to the domestic fresh water system and injecting water, via a spray orifice before the protective grating at turbine inlet, for a period of 15 to 20 minutes with drains open to discharge excessive moisture which does not evaporate off. Since water washing may not completely remove deposits, and can interact with sulphurous deposits with resultant corrosive attack, chemical cleaning may be used in preference. This effectively removes deposits at the turbine and moreover is still active within the exhaust gases passing to the waste heat system, so that further removal of deposits occurs which maintains heat transfer at optimum condition and keeps back-pressure of exhaust system to well within the limits required for efficient engine operation.

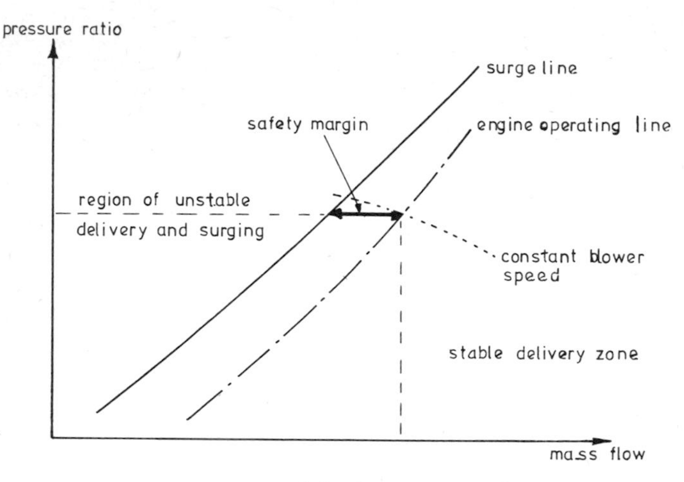

Fig. 56

Surging

With centrifugal compressors the phenomena of surging is primarily due to the mass flow of air at a predetermined pressure ratio falling below a minimum level. To obtain efficient and stable operation of the charging system it is essential that the combined characteristics of the engine and blower are carefully matched. The engine operating line, as indicated on Fig. 56, is mainly a function of these characteristics and taking into account the fact that blower efficiency decreases as the distance between surge and operating lines increases, the matching of blower to engine becomes a compromise between acceptable blower efficiency and a reasonable safety margin against surge. An accepted practice is to provide a safety margin of around 15 to 20° to allow for deterioration of service conditions such as fouling and contamination of turbo-chargers and increasing resistance of ship's hull, etc. Apart from fouling of turbo-charger other contributory factors to surging are contamination of exhaust and scavenge ducting, ports and filters. Since faulty fuel injection leads to poor combustion and greater release of contaminants the need to maintain fuel injection equipment at optimum condition is essential. Other related causes are variation in gas supply to turbo-chargers due to unbalanced output from cylinder units and mechanical damage to turbine blading, nozzles or bearings, etc.

During normal service the build-up of contaminants at the turbo-charger can be attributed to deposition of air-borne contaminants at the compressor which in general are easily removed by waterwashing on a regular basis. At the turbine however, more active contaminants resulting from vanadium and sodium in the fuel together with the products of incomplete combustion deposit at a higher rate which increases with rising temperature. A further problem arises with the use of alkaline cylinder lubricants with the formation of calcium sulphate deposits originating from the alkaline additives in the lubricant. Again water washing on a regular basis is beneficial in removing and controlling deposits but particular care needs to be taken to ensure complete drying out after washing sequence since any remaining moisture will interact with sulphurous compounds in exhaust gas stream with damaging corrosive effect.

Turbo-Charger Breakdown

For correct procedure, depending upon engine type, reference should be made directly to the engine builders and/or turbocharger manufacturers recommended practice. As a general rule however, in the event of damage to the turbo-chargers, the engine should be immediately stopped in order that the damage is limited and does not become progressive. Under conditions where the engine cannot be stopped, without endangering the ship, engine speed should be reduced to a point where turbo-charger revolutions have fallen to a level at which the vibration usually associated with a malfunction is no longer perceptible.

If the engine can be stopped but lack of time does not permit *in situ* repair or possible replacement of defective charger it is essential that the rotor of the damaged unit is locked and completely immobilised. If exhaust gas still flows through the affected unit once the engine is restarted, the coolant flow through the turbine casings needs to be maintained but due to the lack of sealing air at shaft labyrinth glands the lubricating oil supply to the bearings will need to be cut off — with integral pumps mounted on the rotor shaft, the act of locking the shaft ensures this — otherwise contamination of lubricant together with increase in fouling will occur. For rotor and blade cooling a restricted air supply is required and can be achieved by closing a damper or flap valve in the air delivery line from the charger, to a position which gives limited flow from scavenge receiver back to the damaged blower. Alternatively a bank flange incorporating an orifice of fixed diameter can be fitted at the outlet flange of the blower.

With only a single blower out of a number inoperative the power developed by the engine will obviously depend upon charge air pressure attainable. At the same time a careful watch must be kept upon exhaust condition and temperature to ensure efficient engine operation with good fuel combustion. In the event of all turbo-chargers becoming defective it is possible to remove blank covers from the scavenge air receiver so that natural aspiration supplemented by underpiston effect, etc. or parallel auxiliary blower operation is possible — if this method of emergency operation is carried out protective gratings must be fitted in place of blind covers at the scavenge air receiver. In all cases when running at reduced power special care must be taken to ensure any out of balance due to variation in output from affected units does not bring about any undue engine vibration.

TEST EXAMPLES

Second Class

1. Discuss (a) Scavenging by the loop method, (b) Scavenging by the uniflow method. What is meant by scavenge efficiency? What effects would choked ports have on the performance?

2. Sketch and describe the manner in which an exhaust blower delivers the charge of air to the cylinders of a diesel engine. What arrangements are made to cool the charge? How is the water separated from the charge?

3. Sketch the timing diagram of a 2-stroke marine diesel showing the crank angles at essential points on the diagram. Why in some cases is scavenge air allowed to flow into the cylinder after the exhaust closes? Show what arrangements are made to provide for such a flow.

4. Show with the aid of sketches the essential differences between the following methods of scavenge air production: (a) a free running turbo blower, (b) A free running turbo blower combined with a reciprocating scavenge pump, (c) A turbo blower and under piston charging.

5. What difference might be expected in the scantlings of naturally aspirated engines and supercharged engines? Discuss reasons for these differences.

6. Sketch the timing diagram for an opposed piston engine and state why the eccentrics are given lead in the ahead direction. What adverse effects — if any — would ahead lead have on astern operation?

First Class

1. Describe with the aid of sketches the rotor of a turbo-charger. Show how gas and air leakage is avoided. State how the blower can raise scavenge air pressure over and above that of the exhaust gas which is driving it.

2. What are the nature of the deposits left in the internal casings and surfaces of an exhaust gas blower? How are these deposits removed and what effect do they have on the running of the machinery in general?

3. Describe the procedure to be taken in the event of complete failure of a turbo-blower. What steps must be taken to get the vessel under way within a reasonable time? How is the engine affected by cutting out the blower and what precautions must subsequently be taken?

4. Describe the malfunction of turbo-blower with respect to: (a) surging, (b) vibration, (c) cooling, (d) lubrication.

5. Why is it generally considered essential for turbo-blower bearings to be mounted in resilient housings and how is this resilience achieved? Discuss the relevant merits and demerits of both plain and ball or roller bearings.

6. During normal operation of a large marine diesel engine a turbo-charger begins to surge. What investigations should be carried out and what remedial action is needed to alleviate this condition? If surging were allowed to continue what are the possible effects?

7. Both cylinder head poppet valves and exhaust pistons are utilised to control exhaust from cylinders. Sketch typical timing diagrams and comment on both methods. Also explain why, in an exhaust system, the temperature at the exhaust manifold may indicate a higher temperature than the individual exhaust from each cylinder.

CHAPTER 5

STARTING AND REVERSING

Starting air overlap

Some overlap of the timing of starting air valves must be provided so that as one cylinder valve is closing another one is opening. This is essential so as to ensure no angular position of the engine crankshaft with insufficient air turning moment to give a positive start. The usual minimum amount of overlap provided in practice is 15°. Starting air is admitted on the working stroke and the period of opening is governed by practical considerations with three main factors to consider:

1. The Firing Interval of the engine.

$$\text{Firing Interval} = \frac{\text{Number of degrees in engine cycle}}{\text{Number of cylinders}}$$

e.g. with a four cylinder Doxford engine (2-stroke) the firing interval is 90°, *i.e.* 360/4 and if each cylinder valve covered 90° of the cycle then the engine would not start if it had come to rest in the critical position with one valve fractionally off closure and another valve just about to start opening.

2. The valve must close before the exhaust commences. It is rather pointless blowing high pressure air straight to exhaust and it could be dangerous.

3. The cylinder starting air valve should open after firing dead centre to give a positive turning moment in the correct direction. In fact some valves are arranged to start to open as much as 10° before the dead centre because the engine is past this position before the valve is effectively well open, and in fact any reverse turning effect is negligible as the turning moment exerted on a crank very near dead centre is small indeed.

Consider Fig. 57 (a) for a 4-stroke engine. With the timings as shown the air starting valve opens 15° after dead centre and closes 10° before exhaust begins. The air start period is then 125°. Firing interval for a 6-cylinder 4-stroke engine = 720/6 = 120°. The period of overlap is 5° which is insufficient. Although this example could easily be modified so as to give sufficient (say 15°) overlap by reducing the 15° after dead centre and the 10° before exhaust opening, it can become very difficult to arrange with very

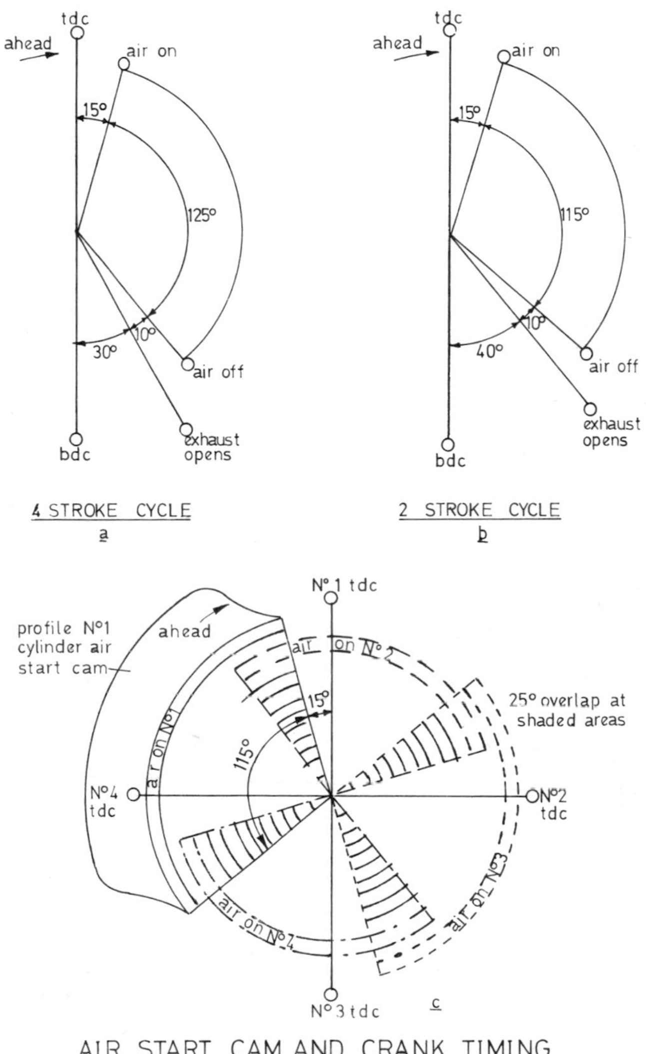

AIR START CAM AND CRANK TIMING DIAGRAMS

Fig. 57

early exhaust opening on turbo-charged engines and a 7-cylinder 4-stroke engine is much easier to arrange.

Consider Fig. 57 (b) for a 2-stroke engine:

This has an air start period of 115°. Firing interval for a 3-cylinder 2-stroke engine = 360/3 = 120°. This means no overlap. Modification can arrange to give satisfactory starting with this example but for modern turbo-charged 2-stroke engines having exhaust opening as early as 75° before bottom (outer) dead centre it becomes virtually impossible. A 4-cylinder 2-stroke engine is much easier to arrange and would be adopted but in fact for modern practice requiring such high powers the number of cylinders is increasing and 4-cylinder engines are becoming in the minority so that there are virtually no air period starting problems.

Consider Fig. 57 (c) which is a cam diagram for a 2-stroke engine with 4 cylinders. The air open period is 15° after dead centre to 130° after dead centre, *i.e.* a period of 115°. This gives 25° of overlap (115 – 360/4) which is most satisfactory. Take care to note the direction of rotation and that this is a cam diagram so that for example No. 1 crank is 15° after dead centre when the cam would arrange to directly or indirectly open the air start valve. The firing sequence for this engine is 1 4 3 2. This is very much related to engine balancing and no hard and fast rules can be laid down about crank firing sequences as each case must be treat on its merits.

It may be useful to note that for 6-cylinder, 2-stroke engines a very common firing sequence is 1 5 3 6 2 4 and similarly for 7- and 8-cylinders 1 7 2 5 4 3 6 and 1 6 4 2 8 3 5 7 respectively are often used.

The cam on No. 1 cylinder is shown for illustration as it would probably be for operating say cam operated valves, obviously the other profiles could be shown for the remaining three cylinders in a similar way. The air period for Nos. 1, 4, 3 and 2 cylinders are shown respectively in full, chain dotted, short dotted and long dotted lines and the overlap is shown shaded.

Starting Air Valves (Air piston operated)

This valve is fitted to the cylinder and is opened by air pressure directed to it from the air distributor and is closed when the air distributor vents the line to atmosphere via a silencer. The valve illustrated is for no particular engine design but almost all modern engines have valves working on identical principles and very similar actual design. The body of the valve could be of mild steel, the spindle of high tensile steel and the steel valve could have the contact faces stellited or hardened.

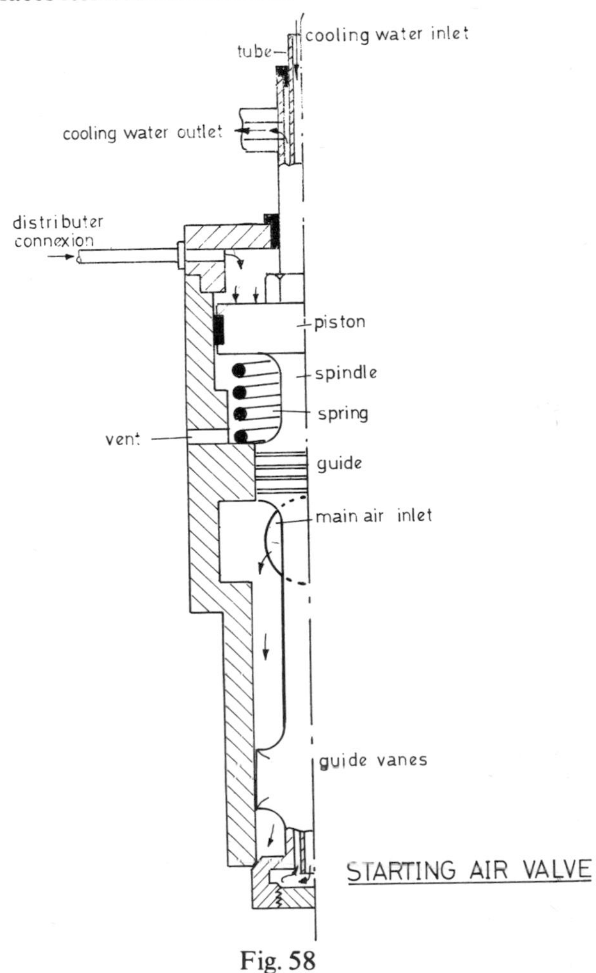

Fig. 58

The valve is shown water cooled but this is not essential. Lubrication is by grease and the normal valve lift is about 20 mm. A non-return safety valve and flame trap would be fitted in the main air supply to the cylinders. Note that in detail sketches the usual practice is to use half views wherever possible in order to save time for examination sketching. The other half of the sketch would be a mirror image. Main air supply is at the back and this sketch should be compared with the more simple and smaller scale drawing used in Fig. 63.

The principle of operation is as follows:

The main air enters and the down force on the valve tending to open it is approximately balanced by the up force on the spindle guide (areas nearly equal). The spring force up ensures that the valve will remain closed provided there is no real pressure above the piston (*i.e.* the distributor is venting that valve). Now if the distributor connects the space above the piston to high pressure air this is sufficient to overcome the spring and move the valve downwards so admitting air to that cylinder. This type of valve is sometimes called relay operated. During manoeuvring all cylinder valves have main air supply acting on them.

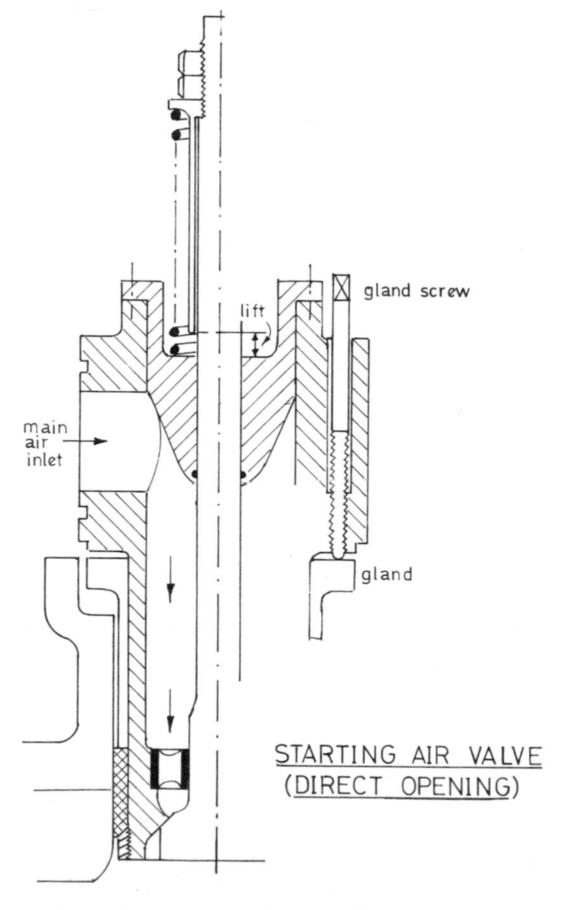

Fig. 59

Starting Air Valve (Direct opening)

Water cooling is not shown on this valve but the arrangement if required could be similar to that for the air piston type valve already shown in Fig. 58. This sketch gives an idea of one method of giving a gas-tight joint into the cylinder liner. The operating principle is simply that opening will occur when main air inlet pressure acting on the valve overcomes the spring force. This type of valve is most generally used where the main air supply is only admitted via a cam operated valve so that air only passes to that cylinder requiring starting air an instant before valve opening is required. This is obviously a complete difference in principle from that of the distributor operated air piston type of relay valve as described previously.

Starting Air Distributor

There are many designs of air distributor all with the same basic principle, *i.e.* to admit air to the pistons of cylinder relay valves in the correct sequence for engine starting. Valves not being supplied with air would be vented to the atmosphere via the distributor. Some overlap of timing would obviously be required.

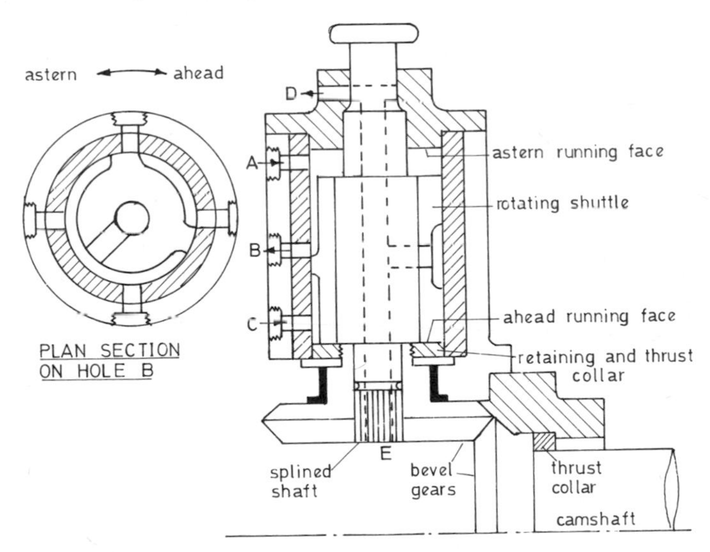

STARTING AIR DISTRIBUTOR

Fig. 60

One type of revolving shuttle design of distributor is as sketched in Fig. 60 this is based on the Doxford type. Another type utilising a cam operated design is shown as part of Fig. 63.

Considering Fig. 60.

The driving gears run in an oil bath and there is about 0.08 mm end float arranged vertically before the retaining collar is locked with a retaining screw. The shuttle is grease lubricated, has a small working clearance of about 0.08 mm and is free to slide on the splined shaft. A pilot valve (not shown) is operated from the main control air start lever and is arranged to either, (a) open a main isolating (automatic) valve to let main air flow to the cylinder air starting valves and also let air flow to Ahead (port A), or (b) open the isolating valve and let air flow to Astern (port C), or (c) vent the system.

The distributor of this particular design is arranged to operate no more than five cylinders. If the air lever is moved to Ahead the distributor air from the pilot valve enters at hole A and passes to the appropriate cylinder air starting valve through the outlet hole B. Which air valve receives air depends on the rotary position of the shuttle. The engine turns and the shuttle closes port B to air and connects it to atmosphere through the hole bored in the centre of the shuttle and ports D and E. If the air lever is moved to Astern the distributor air from the pilot valve enters at C and presses the shuttle up along the splines so that outlet hole B is now connected to Astern air (with different timings) and the appropriate valve. If the air lever is at air off the system is vented and the shuttle is on its bottom face. Obviously plan sections at a number of positions vertically would be required to show a complete detail of the port arrangements. For examination purposes this has been reduced to one plan section on hole B, 4-cylinder connections are shown, two are venting through the shuttle, one is blanked off and the remaining one is venting through the pilot valve as the air start lever will be at air off position.

General Reversing Details

Most main engines are of the direct coupled 2-stroke type, medium or high speed engines require reduction gearing and the trend here is to unidirectional types. Hence the need for reversing mechanisms for 4-stroke engines is reducing. For these reasons the 2-stroke reversing mechanism will be considered in greatest detail.

2-stroke reversing gear

It is usually necessary to reposition the fuel cams on the camshaft, with jerk pumps, so that reversing can utilise one cam this avoids the complication of moving the camshaft axially. This means that it is necessary to provide a lost motion clutch on the camshaft and the need for such a clutch will first be described.

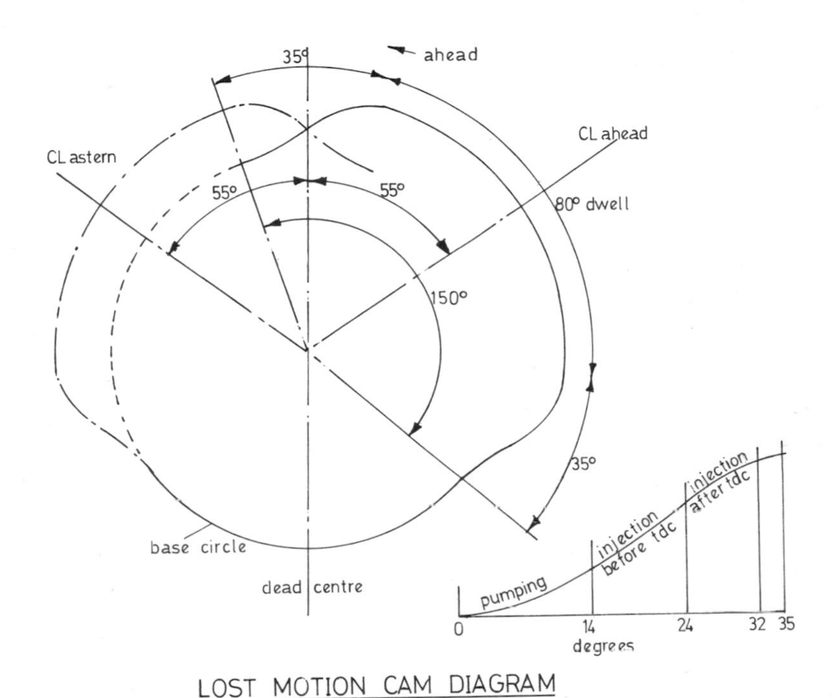

Fig. 61

Referring to Fig. 61 the lost motion cam diagram:

Consider the engine position to be dead centre Ahead with the cam peak centre line to be 55° after this position, anti-clockwise ahead rotation, for correct injection timing ahead. If now the engine is to run astern (clockwise) the cam is 55 + 55 = 110° out of phase. Either the cam itself must be moved by 110° or while the engine rotates 360° the cam must only rotate 250° (110° of lost motion). Note the symmetrical cam 75° each side of the cam peak centre line made up of 35° rising flank and 40° of dwell.

The flank of the cam is shown on an enlarged scale in Fig. 61. It will be noted that the 35° of cam flank is utilised for building up pressure by the pumping action of the rising fuel pump plunger (14°) for delivery at injection 10° before firing dead centre to 8° after firing dead centre, and 3° surplus rise of flank for later surplus spill variation. It is obvious that the lost motion is required with jerk pumps, cam driven, in which a period of pumping is necessary before injection starts. The following points are worth specific mention:

1. Early Doxford engines had a camshaft (front) which moved axially for astern. Later Doxford engines have cam operated valves, not pumps, from the camshaft so no lost motion is required.

2. The dwell period is not normally necessary from the fuel injection aspect alone, *i.e.* about 30° lost motion would be adequate and is provided as such on British Polar and older Sulzer engines.

3. Dwell, in which the fuel plunger is held before return, is often provided to give a delay interval. For example with some B. and W. engines about 80° dwell gives a rotation (total) of the camshaft of about one third of a revolution which allows an axial travel with a screw nut arrangement of reasonable size and pitch to change over chain driven blower air valves for reverse running.

4. Modern Sulzer engines have about 98° lost motion as the distributor repositioning for astern is from the same drive shaft as the fuel pumps, but via a vertical direct drive shaft.

Refer now to Fig. 62 the Lost Motion Clutch.

This design which is based on older Sulzer engine practice has a lost motion on the fuel pump camshaft of about 30°. When reversal is required oil pressure and drain connections are reversed. Oil flowing laterally along the housing moves the centre section to the new position, *i.e.* anticlockwise as shown on the sketch in Fig. 62. The oil pressure is maintained on the clutch during running so that the mating clutch faces are kept firmly in contact with no chatter.

There are a number of variations on this design but the principle of operation is similar although not all types rotate the clutch to its new position before starting and merely allow the camshaft to 'catch on' with the crankshaft rotation when lost motion is completed.

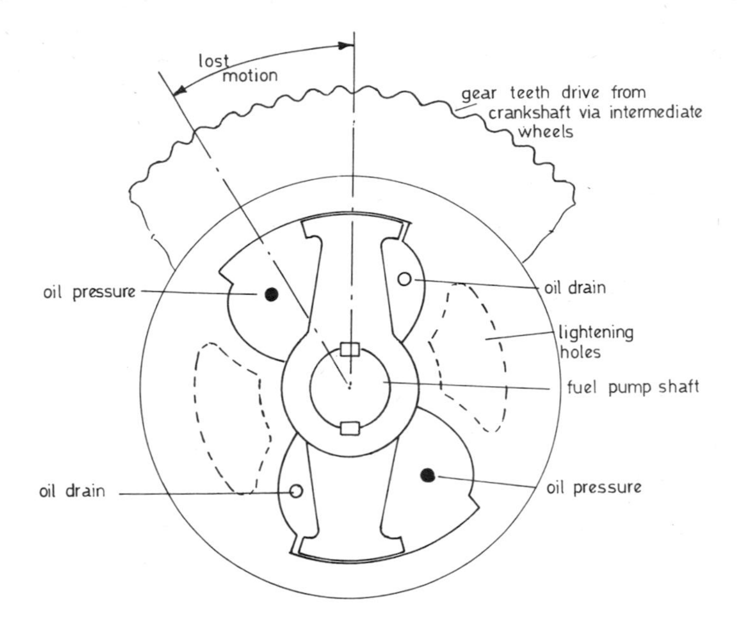

LOST MOTION CLUTCH (early Sulzer)

Fig. 62

PRACTICAL SYSTEMS

Having described the basic principles of starting and reversing the actions are now combined to give a selection of systems as used on the various engine types.

Starting Air System (B. and W.)

Consider first the air off position. Air from the storage bottle passes to the automatic valve which however remains shut as air passes through the pilot valve (1) to the top of the automatic valve piston. All cylinder valves and distributor valves are venting to atmosphere via the automatic valve. If now the lever is moved to the position shown in Fig. 63 then the air pressure on top of the automatic valve is vented through the pilot valve (1) by the linkage shown. This causes the automatic valve to open as the up force on the larger piston is greater than the down force on the smaller valve with the spring force. The lower vent connection is closed

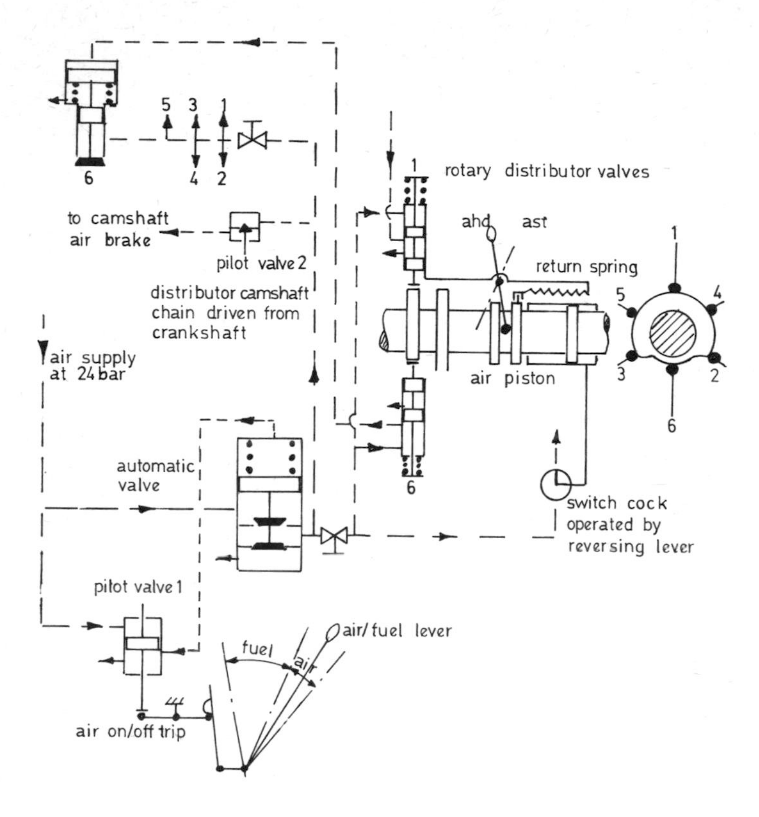

STARTING AIR SYSTEM (B&W)

Fig. 63

and air flows to all cylinder and distributor valves. The cylinder valves are of the air piston relay type described earlier and in spite of main air pressure on them will be closed except for one valve (or possibly two). This distributor has the piston pilot valves mounted around the circumference of a negative cam. Only one distributor pilot valve can be pushed into the negative cam slot, *i.e.* No. 6 and hence air flows through the No. 6 distributor pilot valve only to the upper part of the piston of the No. 6 cylinder air starting valve, which will open. All other starting valves are shut and venting to atmosphere. The position shown for illustration is air on to No. 6 cylinder of a 6-cylinder engine running ahead. When the lever is moved forward on to fuel the whole system is again vented to atmosphere through the automatic valve.

For astern running the reversing lever is moved over which allows air to pass, via a switch cock, to push the light distributor shaft along by means of an air piston (alternatives are scroll, direct linkage, etc.), so putting the astern distributor cam into line with the distributor pilot valves. Distributor pilot valves are kept out by springs during this operation. The air-fuel lever is then operated as previously described for the engine to run astern. Air start timing for the opposed piston 2-stroke engine design, upon which the above system is typical, is 5° before firing dead centre to 108° after firing dead centre (122° after for astern). B. and W. engines also employ a revolving plug type of distributor on some engine designs. Again some types of these engines utilise an air brake on the main camshaft so that air pistons pressure against spring loaded rollers have the air pressure acting on them through the pilot valve (2), operated from the reversing lever, while the lost motion is being travelled by the engine. The main camshaft is therefore kept stationary and just before the lost motion is complete the air pressure is released to atmosphere so releasing the brake. In this chapter no distinction has been made between B. and W. engines and H. and W. engines.

Starting Air System (Sulzer RD)

Refer to Fig. 64.

Air from the starting receiver at 30 bar maximum flows to the pre-starting valve (via the open turning gear blocking valve shown), and directly to the automatic valve. At the automatic valve air passes through the small drilled passage to the back of the piston and this together with the spring keeps this valve shut as the pilot valve is shut with air pressure on top and atmospheric vent below.

If the air starting lever is operated with control interlocks free, the opening of the pre-starting valve allows air to lift the pilot valve, vent the bottom of the automatic valve and cause it to open as shown. This allows air to pass to the cylinder valve via non-return and relief valves and also to the distributor. The distributor will allow air to pass to the appropriate cylinder valve causing it to open due to air pressure on the piston top. In this design when the piston top of the cylinder valve is connected to the atmosphere for venting the bottom of the valve is connected to air pressure this ensures a rapid closing action. The distributor of this engine is very similar in principle to that shown for the B. and W. engine previously except that a positive cam is used by Sulzer. The system described differs from that of the Sulzer RS engine and also from

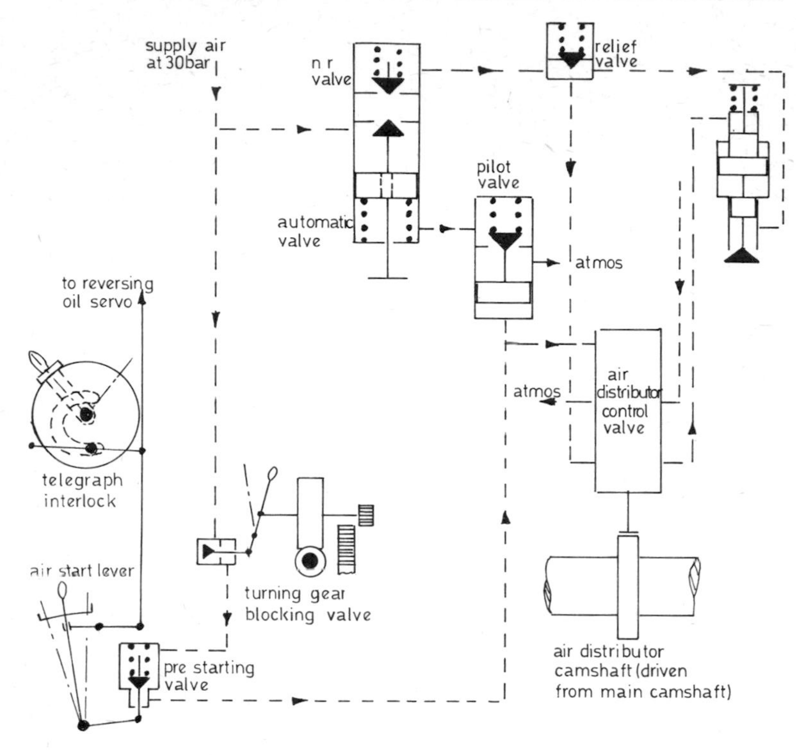

Fig. 64

earlier Sulzer designs, but is selected as probably the most suitable for modern examination purposes. A mechanical interlock is provided as a blocking device from the telegraph as shown. There is also a connection to the reversing oil servo and an interlock connection from the reversing system to the air start lever via a blocking valve. These are described for the next sketch Fig. 65.

Hydraulic Control System (Sulzer RD)

Consider a reversing action from ahead to astern.

Oil pressure from left of reversing valves to right of clutches and under relay valves A and B and air block valve.

The telegraph reply lever on the engine telegraph is first moved to stop and the fuel lever moved back to about notch 3½, the starting lever is mechanically blocked by the linkage shown on the

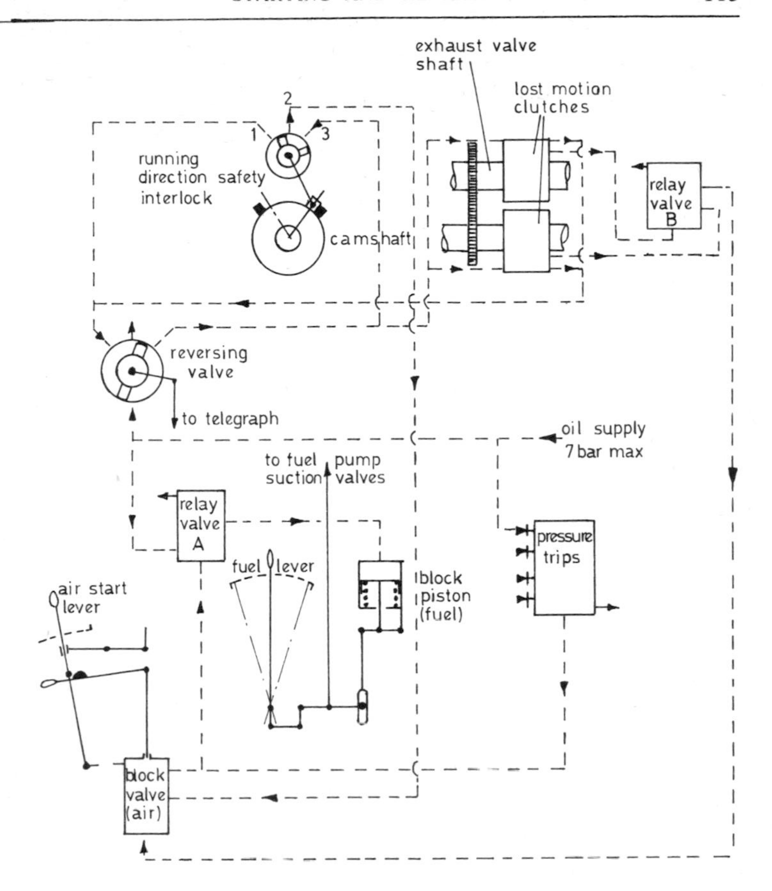

Fig. 65

previous sketch. The telegraph linkage to the reversing valve moves this valve and releases oil pressure from both lost motion clutches. This drop in pressure causes both relay valves A and B to move down by spring action which relieves pressure on the block piston (fuel) so cutting off fuel injection. The pressure on the block valve (cam) is also relieved which serves to also lock the starting lever.

Consider now the situation as shown on the sketch of Fig. 65. When engine speed reduces the telegraph lever can be moved to astern. This allows pressure oil to flow from the right through the reversing valve, as shown on the sketch, to the left of the lost motion clutches to re-position them for astern.

When the servos have almost reached the end of their travel pressure oil is admitted to relay valve B and the block valve (air) releases the lock on the air start lever. (The mechanical lock on the air lever with the telegraph had been released when the telegraph lever was moved to the astern running position.) Pressure oil also acts on relay valve A admitting oil to block piston (fuel) so allowing the fuel control linkages to the fuel pumps to assume a position corresponding to the load setting of the fuel lever.

If the pressure trips act in the event of low oil pressure (supply and bearings) or low water pressure (jacket or piston) then a trip piston moves up under preset spring pressure so connecting the oil pressure connection to drain. This pressure drop causes the block piston (fuel) to rise up under its spring force and shut off fuel injection.

Connections 1 and 3 from the running direction safety interlock to the reversing valve only allow fuel to the engine if the rotation agrees with the telegraph position. If not, the block piston (fuel) is relieved of pressure via block valve (air) and relay valve A.

Movement of the air starting lever can now be carried out as both locks have been cleared and subject to no trip action and satisfactory correspondence between rotation direction and telegraph reply lever indication fuel can be admitted following the full sequence of air starting as described previously, and illustrated in Fig. 64. It is obvious that this system has a large amount of auto-control and is easily adjusted for bridge control.

Control Gear Interlocks

There are many types of safety interlocks on modern IC engine manoeuvring systems. The previous few pages have picked out a number relating to the Sulzer RD engine and these will be adequate to cover most engine type designs as principles are all very similar.

Consider the interlock systems illustrated in Figs. 65 and 66. The telegraph and turning gear interlocks are straight mechanical linkages. In the former case rotation of the telegraph lever from stop position causes the pin to travel in the scroll and unlock the air start lever as well as re-position the reversing valve. The turning gear blocking valve can be seen to close when the pinion is placed in line with the toothed turning gear wheel of the engine. The interlock exerted on the block piston (fuel) is also a fairly simple principle working on the relay valve A from the pressure

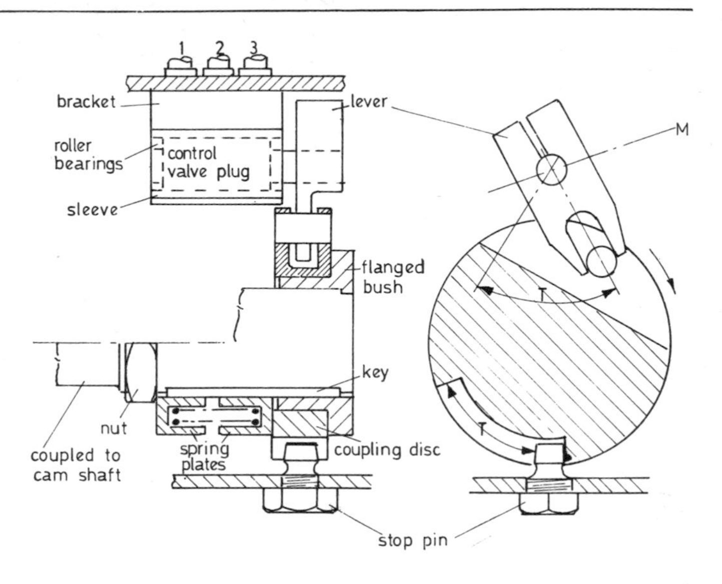

SAFETY LOCK FOR CORRECT ROTATION
(SULZER RD)

Fig. 66

trips and is as described previously. Similarly the block valve (air) operates mechanically via the lever lock on air start lever and horizontal operating lever which rises to unlock under the oil pressure acting through relay valve B on block valve (air) after the clutch reversals have taken place. (The pressure trips are merely spring loaded pistons moving against low oil or water pressure to relieve control oil pressure just like conventional relief valves.) It is perhaps appropriate here to describe one trip in detail and the direction safety lock will now be considered briefly. The function is to withhold fuel supply during manoeuvring if the running direction of the engine is not coincident with the setting of the engine telegraph lever. Refer now to Fig. 66.

At the camshaft forward end the shaft is coupled to the camshaft and carries round with it, due to the key, a flanged bush and spring plates which cause an adjustable friction pressure axially due to the springs and nut. This pressure acts on the coupling disc which rotates through an angular travel T until the stop pin prevents further rotation. This causes angular rotation of a fork lever and the re-positioning of a control valve plug in a new position within the sleeve. Oil pressure from the reversing valve can only pass to the block valve (air) and unlock the air start lever and the fuel control if the rotation of the direction interlock is correct. If the stop pin were to break the fork lever would swing to position M and the fuel supply would be blocked.

Starting Air System (Modern Doxford)

Refer to Fig. 67.

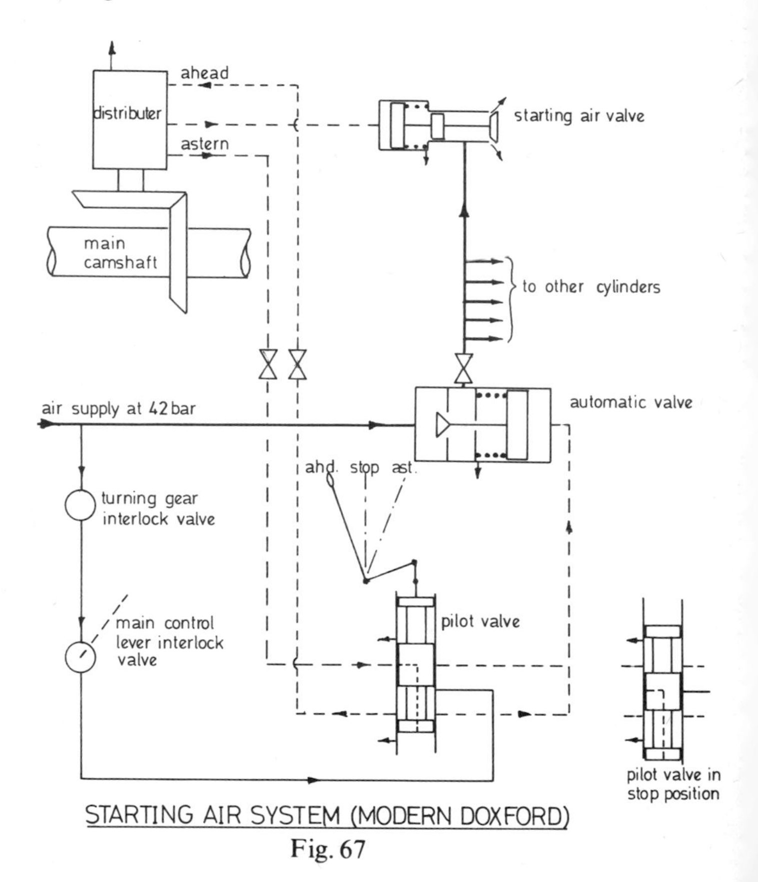

STARTING AIR SYSTEM (MODERN DOXFORD)

Fig. 67

This arrangement should be compared to Fig. 63 as the only real point of difference is the distributor itself, the principle is almost identical. Air from the storage bottle is allowed to reach the pilot valve only if the main control lever is at the off position and the turning gear is clear. Movement of the air lever ahead causes air to pass to the rotary distributor and to the automatic valve via isolating and non-return pilot valves. Movement of the automatic valve right to left occurs against the spring action as the piston area exposed to air pressure on the right is greater than the valve back area exposed to air pressure on the left. The automatic valve then opens fully and air passes to all the starting air cylinder valves (one only shown). Air supply to the cylinder valves is insufficient to open these valves as valve back and piston guide are approximately equal in area, hence up air force balances down air force and the spring maintains the valve shut provided the piston area top is only subject to atmospheric pressure.

Air passes through the distributor to the appropriate valve and acting on the piston top of this valve causes the valve to open and admit air to the cylinder. The case hardened steel shuttle is driven by bevel wheels and circumferential slots machined on it connect with parts on the sleeve body. Vent holes are arranged in the top cover and an axial hole and two radial holes in the shuttle allow venting. When running ahead the lower face of the shuttle is in contact with the body and for astern the shuttle moves up on the splined shaft so that the upper face is in contact with the body. A single distributor does not supply more than five cylinders. A flame trap is fitted between the cylinder valves and the automatic valve.

Starting Air System (Early Doxford)

Refer to Fig. 68.

There is a separate lever for starting air and for fuel, interlock prevents the fuel lever passing notch 2 (notch 10 maximum) with the air lever in the on position. Air pressure from the storage receiver is directed through the pilot valve by operation of the air start lever to the on position (at the off position the atmosphere connection at the pilot valve is open, so venting). Air also passes direct to the cam operated main starting air valves but cannot pass through any one valve until that valve is opened by the cam action. With, for example, a 6-cylinder engine the valves are operated from the front camshaft and are in two blocks together; Nos. 1, 2, 3, cylinder valves forward, and Nos. 4, 5, 6 cylinder valves aft with a relay cylinder for each block.

After movement of the air lever to the on position the pilot valve allows air pressure to move the relay cylinder piston to the right against a spring. With fulcrum of a first lever at A the second lever is drawn in at C with B as fulcrum, this brings the roller to the cam

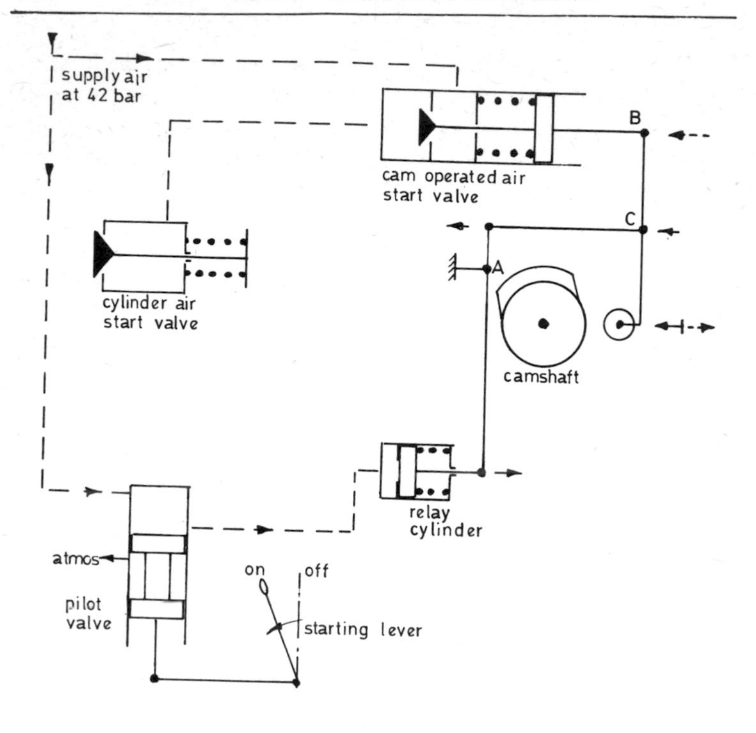

Fig. 68.

base circle (clearance 2 mm). When the cam rotates the toe causes one particular second lever roller to move out left to right with C as fulcrum and hence B moves right to left. This allows main air to flow direct to the appropriate cylinder starting air valve. Both these valves are normally spring shut. The non-return cylinder valve is water cooled and the lift of the main valve is 16.5 mm. There is nothing worthy of particular note about the cylinder valve or the rather complex lever action, etc. of the main valve and as these details refer to old practice no further description is necessary.

TEST EXAMPLES

Second Class

1. Explain the purpose of the "lost motion" provided for the camshaft on certain types of main diesel engines and sketch the arrangement.

2. Describe the reversing gear for a 2-stroke main engine. Proceed through the operation in its proper sequence step by step.

3. Sketch and describe an air starting valve suitable for a main propulsion unit. How can an air starting valve be tested? What may happen if an air starting valve sticks open?

4. With the aid of a sketch describe the interlocking gear in the air starting system of a heavy oil engine, *i.e.* how are the starting and fuel arrangements locked when the telegraph is answered wrongly?

5. Describe, with the aid of a sketch, how a large heavy oil engine is reversed.

6. Describe how you would verify the timing of the air starting valves of a large marine oil engine. What is meant by "starting air overlap" and what is its object?

First Class

1. Sketch and describe an air starting system where the air starting valve opening relative to the crankshaft is controlled by a rotary distributor.

2. Describe, with the aid of sketches, the means of reversing a heavy oil engine by changing the angle of camshaft in relation to crankshaft.

3. Explain how a Sulzer engine is started and reversed. Sketch the main details of the starting and reversing system and pay particular attention to lost motion devices.

4. Describe an air start valve and the operation of an air starting system. Name the engine to which you make reference. Discuss the dangers of a leaky valve and describe the tests carried out on these valves. State suitable air starting pressures giving maximum and minimum values.

5. Describe, with the aid of sketches, how a main engine:
(a) Can only run in the direction appointed by the controls.
(b) Can only run in the direction indicated by the telegraph.
(c) Can run ahead on the astern cams.

6. Sketch, and describe in detail, an air starting system for a heavy oil engine in which the air timing control is remote from the cylinder head. State the type and make of the engine.

CHAPTER 6

CONTROL

The study and application of instrument and control devices has developed from the beginning of engineering itself. It is however in the last few years that this branch of engineering has assumed greater importance. Automatic Control in a simple sense has always been utilised, *e.g.* cylinder relief valves, speed governors, overspeed trips, etc. It is intended in this chapter to

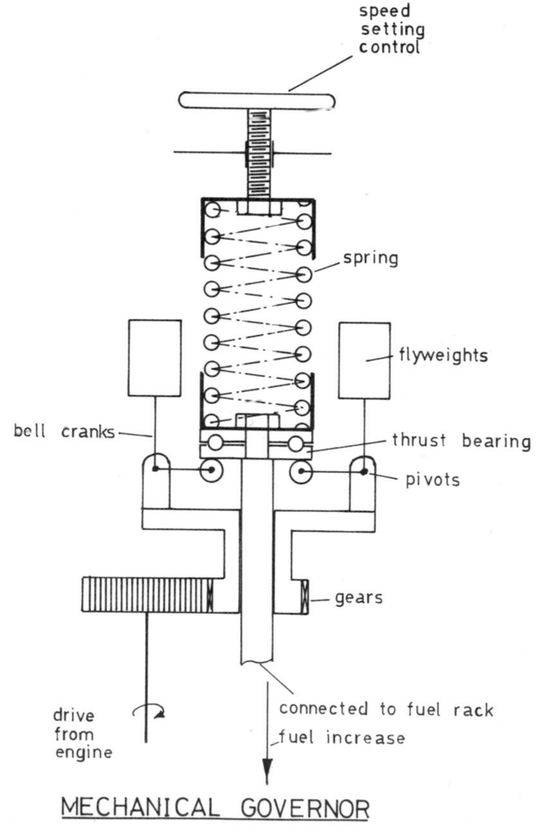

Fig. 69

examine the control of the modern diesel engine and its associated equipment and to apply control terminology, with explanations, where required.

GOVERNING OF MARINE DIESEL ENGINES

A clear distinction is necessary between the function of a governor and an overspeed trip. Diesels driving electrical generators invariably utilised plain flyweight governors of the Hartnell type (Fig. 69), a change in speed resulted in variation of the position of the flyweights and alteration of fuel supply. Larger, slow running, direct drive diesel engines were not generally fitted with such a governor but they invariably were fitted with an overspeed trip, usually of the Aspinal inertia type. This trip was arranged to allow full energy supply under normal operating conditions but in the event of revolutions rising about 5% above normal the energy was totally shut off until revolutions dropped to normal again. At about 15% above normal revolutions the trip would stay locked, with energy shut off, and this would continue until re-set by hand.

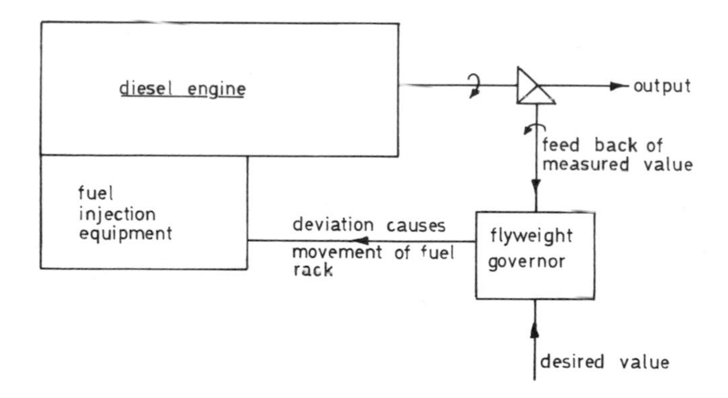

CLOSED LOOP CONTROL

Fig. 70

A plain flyweight governor, which has to perform two separate functions (1) to act as a speed measuring device, (2) to supply the necessary power to move the fuel control lever, is a proportional controller in a closed loop controlled system. Fig. 70 shows in block diagram form the arrangement. A closed loop control system is one in which the control action is dependent on the output. The measured value of the output, in this case the engine speed, is fed back to the controller which compares this value with the desired value of speed. If there is any deviation between the values, measured and desired, the controller produces an output which is a function of the deviation. In this case the controller output would be proportional to the deviation, *i.e.* proportional control.

In control terminology deviation is sometimes called error, since it is the difference between measured and desired values, and desired value is sometimes called set value. Proportional control suffers from offset. In the example, if a speed change occurs the flyweights take up a new equilibrium position and the fuel supply will be altered to suit the new conditions. However the diesel is now running at a slightly different speed to before. If the original speed was the desired value then the new speed is offset from the desired value.

In governor parlance the term speed droop, or just droop, is used to define the change in speed between no load and full load conditions. If speed droop did not exist then there would be one speed only for any position of the governor flyweights and this in turn means any fuel supply rate. In this case the diesel would hunt.

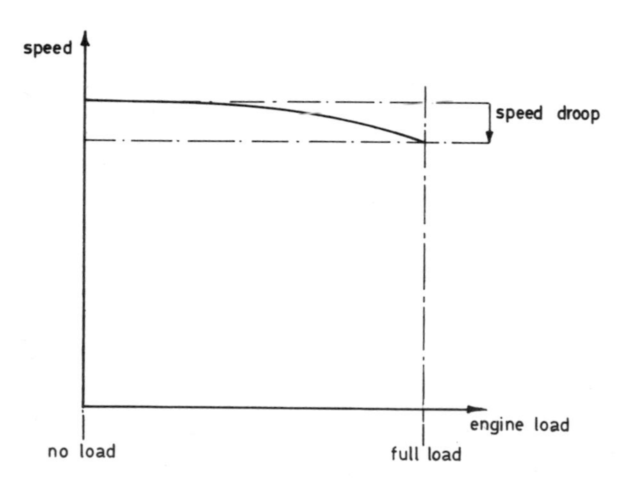

SPEED DROOP

Fig. 71

This is an isochronous condition, an engine fitted with an isochronous governor will hunt. However, the term isochronous has taken on a new meaning as we will find later.

Forces involved in the flyweight governor movement are, inertia, friction and spring. Considerable effort may therefore be required to cause movement, this would necessitate a change in speed without any alteration in governor position. This is bad control, the system is insensitive and various equilibrium speeds are possible. For simple systems these various equilibrium speeds are not an embarrassment, but if we require a better controlled system the two functions that the flyweight governor has to perform would be separated into (1) a speed measuring device and (2) a servo-power amplifier.

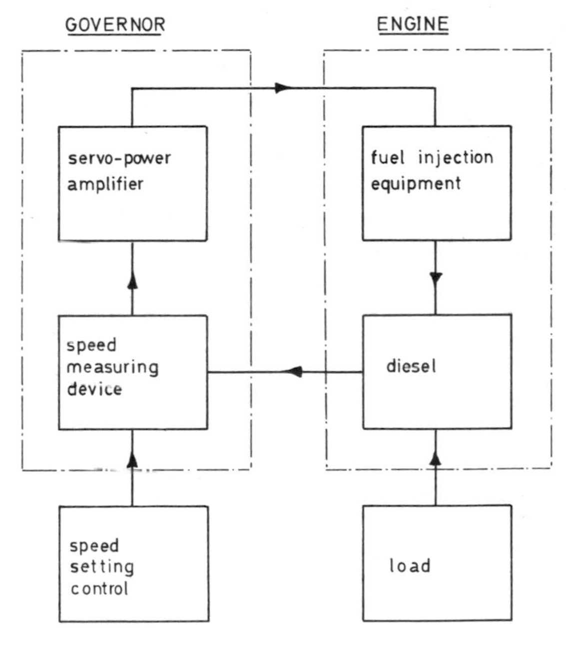

Fig. 72

Fig. 72 shows the basic arrangement in block form. A load increase would cause a momentary speed droop. The speed measuring device would obtain a measured value signal from the diesel and compare this with a desired value from the speed setting control. The deviation would be converted into an output that would bring into action the servo-power amplifier which would position the fuel rack, increasing the supply of fuel to meet the increase in load.

Since the speed measuring device does not have to position the fuel rack — in fact it could be near zero loaded — it can be very responsive, minimising the time delay between load alteration and fuel alteration in the closed loop. The servo-power amplifier is usually a hydraulic device that simply, quickly and effectively provides the necessary muscle to move the fuel rack.

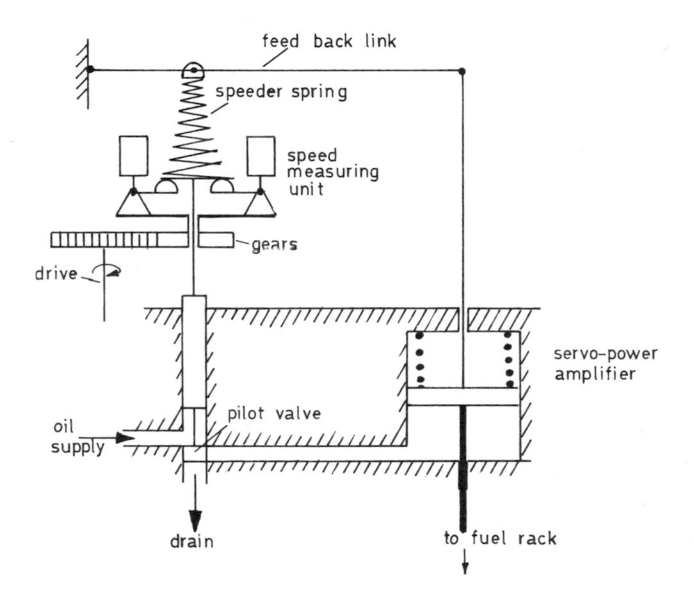

PROPORTIONAL ACTION GOVERNOR

Fig. 73

A proportional action governor is diagrammatically shown in Fig. 73. The centrifugal speed measuring unit is fitted with a conically shaped spring, unlike that shown in Fig. 69, this gives a spring rate which varies as the square of the speed. This gives linearity to the speed measuring system, *i.e.* the response is directly proportional to the change in speed. If we consider an increase in load on the engine, the pilot valve will move down due to the speed drop. The piston in the servo-amplifier will move up and increase the fuel supply to the engine. The feedback link

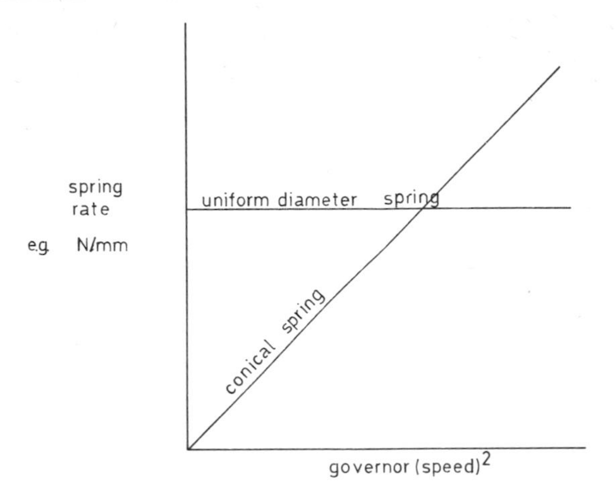

Fig. 74

reduces the force in the speeder spring so that the flyweights can move outwards to a new position, thus raising the pilot valve and closing off the oil supply. If for some reason the oil supply system should fail then the spring loaded piston in the servo-cylinder would be moved down and fuel to the engine cut off. This is called fail safe. Any oil that leaks past the servo-piston will be drained off to the oil sump tank, if this were not so the servo-piston would eventually lock in position.

Flywheels and their effect

Flywheel dimensions are dictated by allowable speed variation due to non-uniform torque caused by individual cylinders firing, this of course is outside the control of the governor. If the speed has to remain nearly constant during changes of load it may be decided to fit a large flywheel, this increases the moment of inertia of the system and gives an integral effect — this must not be taken to extremes or instability may occur. Flywheels however are not cheap and a less expensive solution to the problem may be to fit a better governor.

Integral effect or as it is often called 'reset action' reduces offset to zero, *i.e.* during load alteration the speed will go from the desired value but the reset action works to return the speed to the desired value, so that after the load change the speed is the same as before.

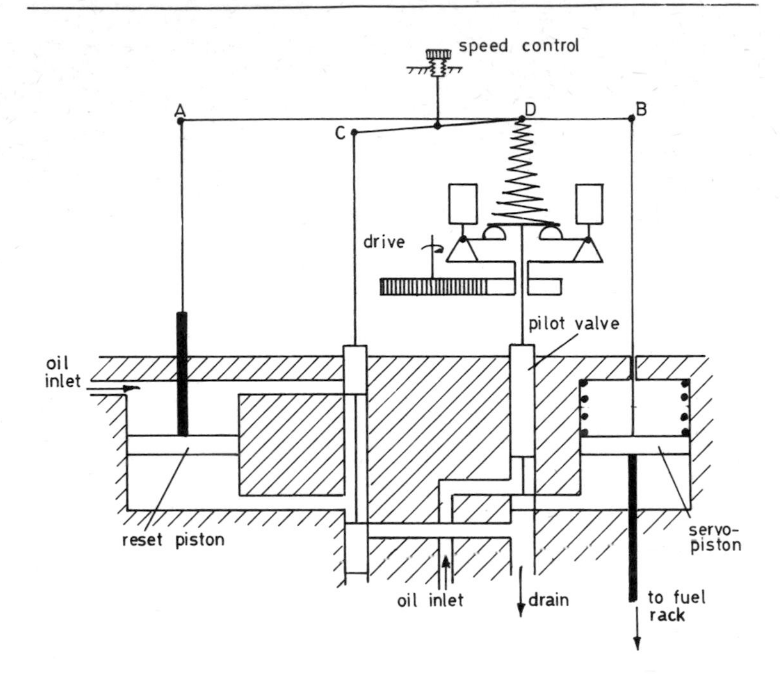

GOVERNOR WITH PROPORTIONAL AND RESET ACTION

Fig. 75

Governor with proportional and reset action

Fig. 75 shows diagrammatically the type of governor that will, after an alteration in engine load, return the speed of the engine back to the value it was operating at before the alteration. If an increase in engine load is considered, the flyweights will move radially inwards and the pilot valve will open to admit oil to the servo-piston. The servo-piston will move up the cylinder compressing the spring and at the same time it will cause (1) the fuel rack to be repositioned to increase fuel supply to the engine, (2) rotate the feedback link 'A-B' anti-clockwise about the pivot point 'A' (this point 'A' would initially be locked due to equal pressures on either side of the reset piston), (3) rotate link 'C-D' anti-clockwise. This rotation of the link 'C-D' will move the reset piston control valve down and some oil will drain from the reset piston cylinder. As the reset piston moves down to a new equilibrium position the feedback link 'A-B' will pivot about 'B' and the link 'C-D' will be rotated clockwise, closing the drain from the reset piston cylinder (and thus locking the reset piston in a new position, returning the point 'D' to it's original position. This means that the engine is now running at it's original speed but with increased fuel supply. Speed droop that took place during the change of the relative positions of the two pistons was transient. This type of governor, that has proportional and reset action, is called in governor parlance an 'isochronous governor.'

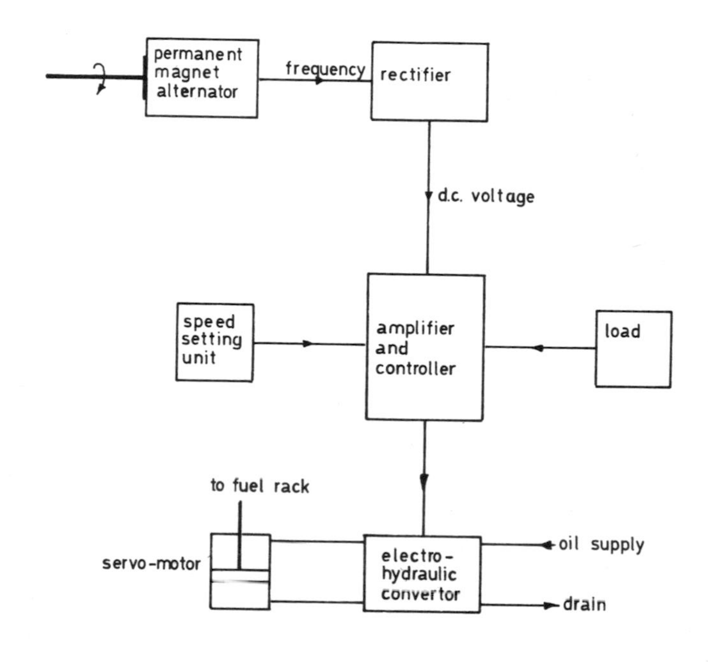

ELECTRIC GOVERNOR

Fig. 76

Electric Governor

This governor has proportional and reset action with the addition of load sensing. A small permanent magnet alternator is used to obtain the speed signal, the advantage to be gained is that there will be no slip rings or brushes with their attendant wear. The speed signal obtained from the frequency of the generated a.c. voltage impulses is converted into d.c. voltage which is proportional to the speed. A reference d.c. voltage of opposite polarity, which is representative of the desired operating speed, is fed into the controller from the speed setting unit. These two voltages are connected to the input of a electronic amplifier. If the two voltages are equal and opposite, they cancel and there will be no change in amplifier voltage output. If they are different, then the amplifier will send a signal through the controller to the electro-hydraulic convertor which will in turn, via the servo-motor, reposition the fuel rack. In order that the system be isochronous the amplifier-controller has internal feedback.

Load Sensing

The purpose of including load sensing into the governor is to correct the fuel supply to the prime mover before a speed change occurs. Load sensing governors are therefore anticipatory governors, *i.e.* they anticipate a change in speed and take steps to prevent, as far as possible, its occurrence.

Load sensing could be done by mechanical means but it would be a complicated and relatively costly system. For this reason it is normally limited to electrical generator drives, it must be remembered that the speed of response of the load sensing element must be better than that of the speed sensing element, which would be used to correct small errors of fuel rack position.

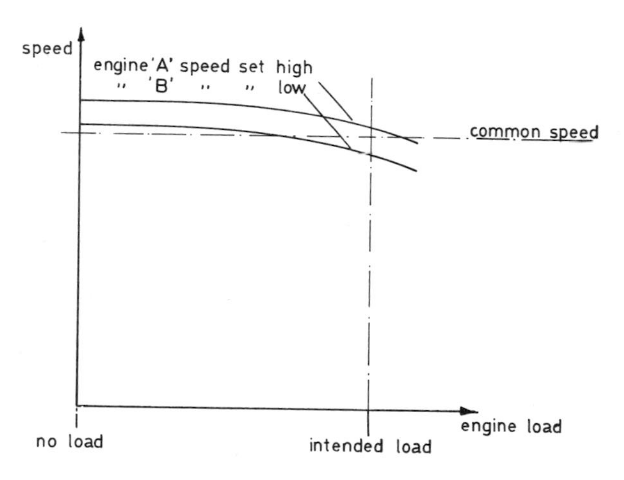

LOAD SHARING BETWEEN TWO ENGINES

Fig. 77

In the example of the electric governor, the electrical output of the main generator would be tapped and if a load alteration took place on the main generator this would be sensed and a signal fed into the controller.

Geared Diesels

Two diesels geared together must run at the same speed, but if the governors of the two are not set equally then they will not carry equal shares of the load.

In Fig. 77 is shown the governor droop curves for two diesels A and B. Governor A has a higher speed setting than that of B, but since they must both run at a common speed the load carried by A will be greater that that of B. Actual loads carried is given by the intersection of the common speed line and the droop curves. By adjusting the speeder settings both droop curves could be made to coincide at the intended load, although this would be difficult to achieve in practice.

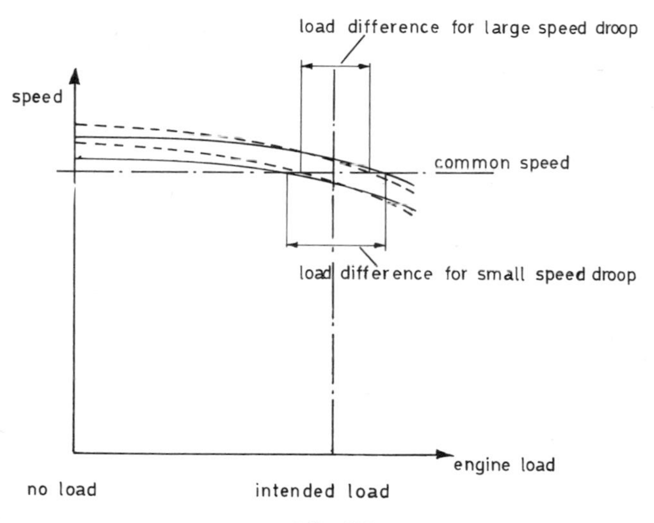

Fig. 78

Shown in Fig. 78 are two sets of droop curves with the same difference in speed setting but with different amounts of speed droop. The difference in load sharing at the common speed is less for the larger speed droop curves than for the smaller. Hence speed droop and fine control over the desired level of speed are necessary for effective load sharing.

Bridge Control of Direct Drive Diesel Engine

Two consoles would be provided, one on the bridge the other in the engine room. For the bridge console the minimum possible alarms and instruments would be provided commensurate with safety and information requirements, *e.g.* low starting air pressure and temperature, sufficient fuel oil, fuel oil pressure and temperature, etc. The engine room console would give comprehensive coverage and overriding control over that of the bridge.

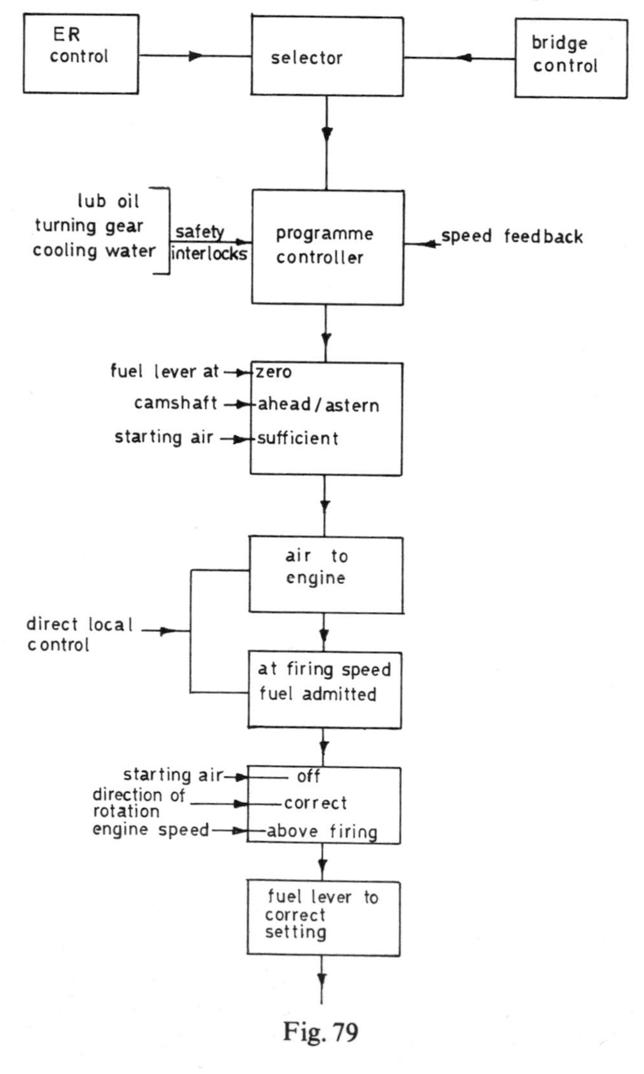

Fig. 79

In Fig. 79 for simplification all normal protective devices are assumed and subsidiary control loops are not considered. The selector would be in the engine room console and the operator can select either engine room or bridge control, with one selected the other is in-operative. Assuming bridge control a programme would be selected, say half ahead. Then providing all safety blockages such as no action with turning gear in, etc. are satisfied, the programme can be initiated and could follow a sequence of checks and operations such as:—

1. Fuel control lever at zero.
2. Camshaft in ahead position.
3. Sufficient starting air.
4. Starting air admitted.
5. Adjustable time delay permits engine to reach firing speed.
6. Fuel admitted.
7. Starting air off, checks on direction of rotation and speed.
8. Fuel adjusted to set value.

Essential safety locks, such as low lubricating oil pressure or cooling water pressure override the programme and will stop the engine at the same time as they give warning.

Direct local control at the engine itself can be used if required or in the event of an emergency.

Further protective considerations:—

1. Governor, including overspeed trip.
2. Non operation of air lever during direction alteration.
3. Failure to fire requires alarm indication and sequence repeat with a maximum of say four consecutive attempts before overall lock.
4. Movement of control lever for fuel for a speed out of a critical speed range if the bridge speed selection is within this range.
5. Emergency full ahead to full astern, etc., actions, must have time delays to allow fall of speed below firing revolutions, astern air open, engine stop, correct astern timing and setting.

Outline Description

The following is a brief description of one type of electronic-pneumatic bridge control for a given large single screw direct coupled I.C. engine to illustrate the main essentials. The I.C. engine lends itself to remote control more easily than turbine machinery.

Movement of the telegraph lever actuates a variable transformer so giving signals to the engine room electronic controller which transmits, in the correct sequence, a signal series to operate solenoid valves at the engine. One set of solenoid valves controls starting air to the engine while a second set regulates fuel supply, the latter via the manual fuel admission lever, is coupled to a pneumatic cylinder whose speed of travel is governed by an integral hydraulic cylinder in which rate of oil displacement is governed by flow regulators. This cylinder also actuates a variable transformer giving a reset signal when fuel lever position matches telegraph setting.

With the engine on bridge control the engine control box starting air lever is ineffective and the fuel control rack is held clear of the box fuel lever. Normal fuel pump control is eliminated and fuel pressure, in a common rail, is automatically adjusted to speed and load by a spring loaded relief valve. Engine override of bridge control is provided.

The function of the electronic controller is to give the following sequence for, say, start to half ahead: Ensure fuel at zero, admit

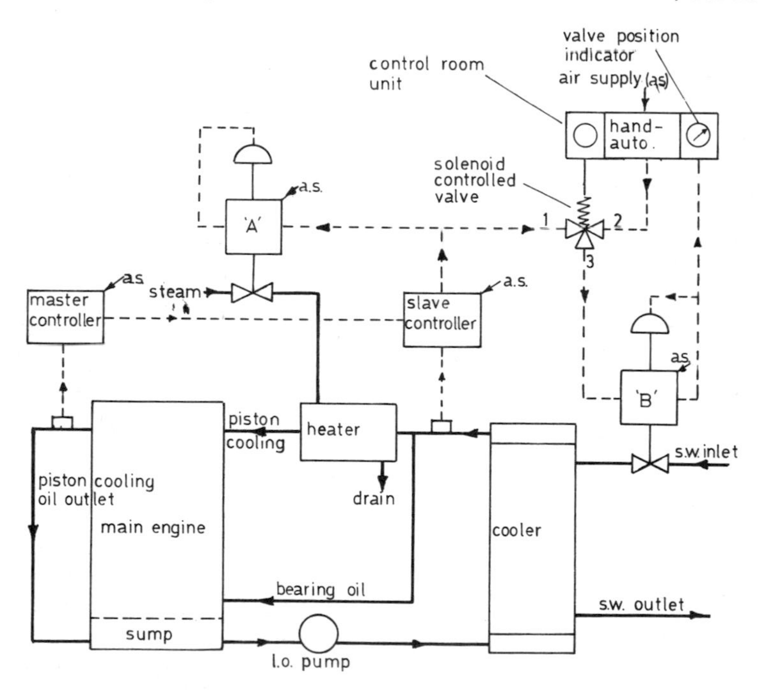

COOLING AND LUB. OIL CONTROL

Fig. 80

starting air in correct direction, check direction, time delay to allow engine to reach firing speed, admit fuel, time delay to cut off air, time delay and check revolutions, adjust revolutions. Similar functions apply for astern or movements from ahead to astern directly. Lever travel time to full can be varied from stop to full between adjustable time limits of ½ minute and 6 minutes. Fault and alarm circuits and protection are built into the system.

PISTON COOLING AND LUBRICATING OIL CONTROL

Simple single element control loops can be used for most of the diesel engine auxiliary supply and cooling loops, however during the manoeuvring of diesel engines considerable thermal changes take place with variable time lags which the single element control may not be able to cope with effectively. (N.B. a single element control system is one in which there is only one measuring element feeding information back to the controller.)

For piston cooling and lubricating oil control the use of a cascade control system caters effectively for manoeuvring and steady state conditions. Cascade control means that one controller (the master) is being used to adjust automatically as required the set value of another controller (the slave).

In Fig. 80 thc two main variables to consider are sea water inlet temperature and engine thermal load. For simplicity we can consider each variable separately:

1. Assuming the engine thermal load is constant and the sea water temperature varies. The slave controller senses the change in lub. oil outlet temperature from the cooler and compares this with its sct value, it then sends a signal to the valve positioner 'B' to alter the sea water flow.

2. Assuming the sea water temperature is constant and the engine thermal load falls. The master controller senses a fall in piston cooling oil outlet temperature and compares this with its set value. It then sends a signal, which is a function of the deviation, to alter the set value of the slave controller. The slave controller now sends a signal to the valve positioner 'B' so that the salt water flow will be reduced and the lub. oil temperature at inlet to the piston increased.

If the engine thermal load is low or zero then valve positioner 'A' will receive a signal from the slave controller which will cause steam to be supplied to the lub. oil heater. This means that the slave control is split between valve positioners 'A' and 'B' — this is called 'split range control' or 'split level control'.

Slave controller output range is 1.2 to 2.0 bar.
Valve positioner 'A' works on the range1.2 to 1.4 bar.
Valve positioner 'B' works on the range 1.4 to 2.0 bar.
Hence the range is split in the ratio 1:3.

Since the piston cooling oil outlet temperature could be offset from the desired value by upwards of 8°C or more, the master controller must give proportional and reset action. In order to limit the variety of spares that must be carried the slave controller would be identical to the master controller.

It may be necessary to change over from automatic to remote control. This is achieved by position control of the three way solenoid operated valve and regulation of the air supply to the

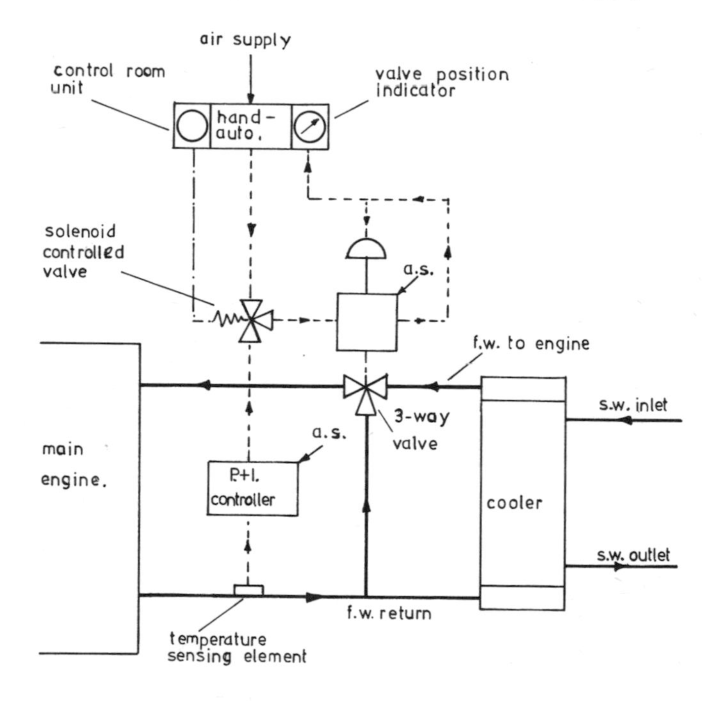

JACKET (OR PISTON) TEMPERATURE CONTROL

Fig. 81

valve positioner 'B' at the control room unit. The solenoid operated valve would be positioned to communicate air lines 2 and 3, closing off 1. Hand regulation of the supply air pressure to valve positioner 'B' enables the operator to control the sea water flow to the cooler. Position of the sea water inlet control valve is fed back to control room unit. Lubricating oil temperatures would be indicated on the console in the control room.

An alternative and often preferred arrangement, using a single measuring element, is to have full flow of sea water through the cooler and operate a three-way valve (2 inlets, 1 outlet) in the

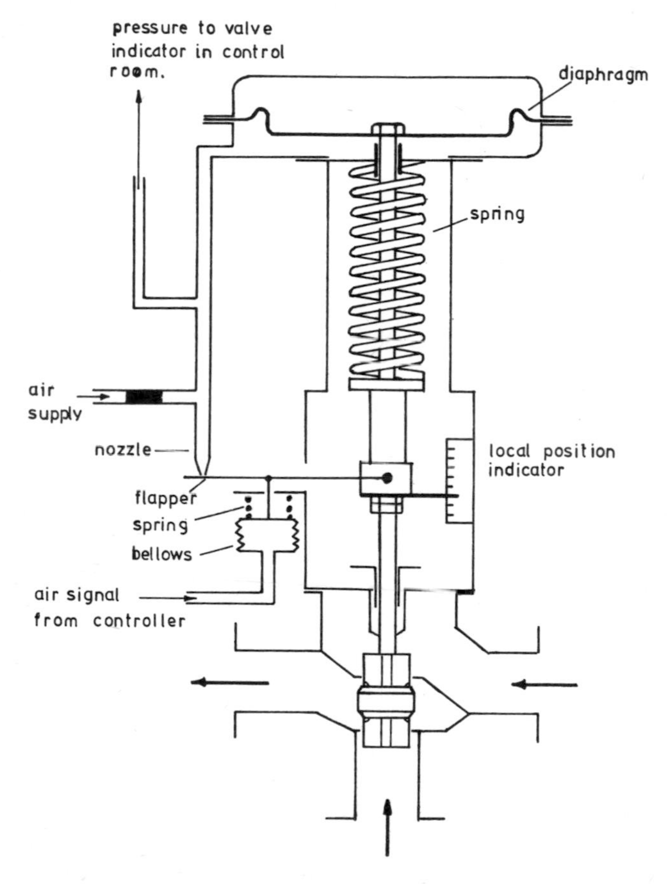

3-WAY VALVE AND POSITIONER

Fig. 82

engine fresh water, or oil, cooling circuit that by-passes the cooler. Valve selection for such duties is most important. Maximum pressure and temperature, maximum and minimum flow rate, valve and line pressure drops, etc., must be carefully assessed so that valve selection gives the best results. With correct analysis of the plant parameters and careful valve selection, simple single element control systems can be employed. This would avoid the extra cost of sophisticated control loops and their attendant increased maintenance and fall in reliability.

For mixing and by-pass operations a three-way automatically controlled valve with two inlets and one outlet of the type shown diagrammatically in Fig. 82 could be used.

An increase in controller output pressure p causes the flapper to reduce outflow of air from the nozzle, the pressure on the underside of the diaphragm increases and the valve moves up. As the valve moves, the flapper will be moved to increase outflow of air from the nozzle and eventually the valve will come to rest in a new equilibrium position.

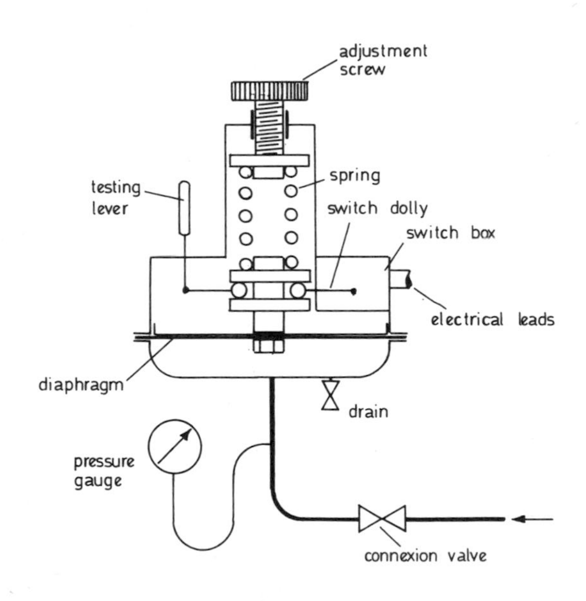

Fig. 83

Indication of valve position is given locally and remote, in the latter case by feeding back the diaphragm loading pressure to an indicator possibly situated in the control room. The valve positioner gives accurate positioning of the valve and provides the necessary muscle to operate the valve against the various forces.

Pressure Alarm

The alarm diagrammatically shown in Fig. 83 can be used for either high or low pressure warning. It can also be used for high or low level alarm of fluids in tanks since pressure is a function of head in the tank.

To test the electrical circuitry and freedom of movement of the diaphragm and switches, the hand testing lever can be used. Setting is achieved, for low pressure alarm, by closing the connection valve and opening the drain. When the desired pressure is reached, as indicated on the gauge, the alarm should sound. If high pressure alarm is required the unit can be set by closing the connection valve and coupling a hydraulic pump to the drain connection.

Unattended Machinery Spaces

These are designated u.m.s. in regulations and in the case of the diesel engines they are gradually increasing in number. A controllable pitch propeller driven by geared unidirectional medium or high speed diesels is a relatively uncomplicated system that lends itself to direct control from the bridge. However, irrespective of the type of installation, certain essential requirements for u.m.s. particularly unmammned engine rooms at night, must be fulfilled. They could be summarised as follows:

1. *Bridge control of propulsion machinery*

The bridge watchkeeper must be able to take emergency engine control action. Control and instrumentation must be as simple as possible.

2. *Centralised control and instruments are required in machinery space*

Engineers may be called to the machinery space in emergency and controls must be easily reached and fully comprehensive.

3. *Automatic fire detection system*

Alarm and detection system must operate very rapidly. Numerous well sited and quick response detectors (sensors) must be fitted.

4. *Fire extinguishing system*

In addition to conventional hand extinguishers a control fire station remote from the machinery space is essential. The station must give control of emergency pumps, generators, valves, ventilators, extinguishing media, etc.

5. *Alarm system*

A comprehensive machinery alarm system must be provided for control and accommodation areas.

6. *Automatic bilge high level fluid alarms and pumping units*

Sensing devices in bilges with alarms and hand or automatic pump cut-in devices must be provided.

7. *Automatic start emergency generator*

Such a generator is best connected to separate emergency bus bars. The primary function is to give protection from electrical blackout conditions.

8. *Local hand control of essential machinery.*

9. *Adequate settling tank storage capacity.*

10. *Regular testing and maintenance of instrumentation.*

TEST EXAMPLES

Second Class

1. Sketch and describe a governor suitable for an auxiliary engine. Explain how it would be set at a desired speed.

2. Discuss the functions of a flywheel and a speed sensing governor. What is meant by speed droop?

3. Explain with reference to a diesel engine the following control terms, (a) closed loop, (b) feed back, (c) desired value.

4. Make a simple line diagram of a jacket or piston cooling system which has automatic temperature control.

5. Sketch and describe an alarm suitable for giving audible warning of low lubricating oil pressure.

First Class

1. Sketch and describe: (a) an automatic temperature control system suitable for the jacket cooling of a diesel engine, (b) an audible alarm device operated by low lubricating oil pressure.

2. Discuss the problems of safety which arise in vessels with automated engine rooms that are unmanned for extended periods.

3. Describe the action of a diesel engine governor when change of engine load takes place. Correlate the following terms to the governor action, (a) sensing element, (b) desired value, (c) error signal, (d) feed back.

4. Sketch and describe a diesel engine generator governor which senses load and speed. Why is this type of governor used in preference to one which senses speed only.

5. Sketch and describe a system of bridge control for a main diesel engine and explain how you would change over to manual control in the event of failure of the remote control system.

6. Sketch and describe a type of valve suitable for the automatic control of temperature in the fresh water cooling system of a diesel engine.

CHAPTER 7

ANCILLARY SUPPLY SYSTEMS

AIR

Compressed air is used for starting main and auxiliary diesels, operating whistles or typhons, testing pipe lines (*e.g.* CO_2 fire extinguishing system) and for workshop services. The latter could include, pneumatic tools and cleaning lances, etc.

Air is composed of mainly 23% Oxygen, 77% Nitrogen by mass and since these are near perfect gases a mixture of them will behave as a near perfect gas, following Boyle's and Charle's laws. When air is compressed its temperature and pressure will increase as its volume is reduced.

Isothermal compression of a gas is compression at constant temperature, this would mean in practice that as the gas is compressed heat would have to be taken from the gas at the same rate as it is being received. This would necessitate a very slow moving piston in a well cooled small bore cylinder.

Adiabatic compression of a gas is compression under constant enthalpy conditions, *i.e.* no heat is given to or taken from the gas through the cylinder walls and all the work done in compressing the gas is stored within it.

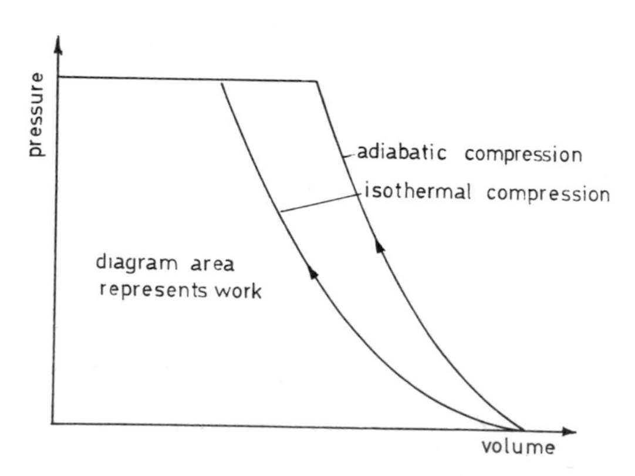

Fig. 84

In Fig. 84 the two compression curves show clearly the extra work done by compressing adiabatically, hence it would be more sensible to compress isothermally. In practice this presents a problem — if the compressor were slow running with a small bore perfectly cooled cylinder and a long stroke piston the air delivery rate would be very low.

Multi-stage compression

If we had an infinite number of stages of compression with coolers in between each stage returning the air to ambient temperature, then we would be able to compress over the desired range under near isothermal conditions. This of course is impracticable so two or three stage compression with interstage and cylinder cooling is generally used when relatively high pressures have to be reached.

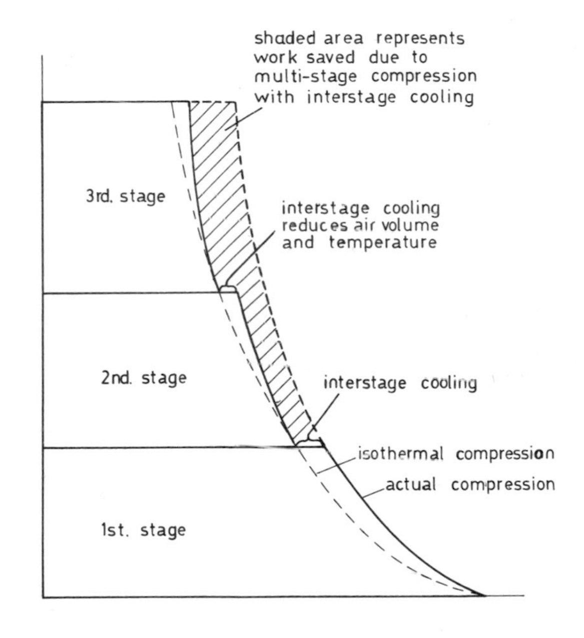

Fig. 85

Fig. 85 shows clearly the work saved by using this method of air compression, but even with efficient cylinder cooling the compression curve is nearer the adiabatic than the isothermal and the faster the delivery rate the more this will be so.

To prevent damage, cylinders have to be water or air cooled and clearance must be provided between piston and cylinder head. This clearance must be as small as practicable.

High pressure air remaining in the cylinder after compression and delivery will expand on the return stroke of the piston. This expanding air must fall to a pressure below that in the suction manifold before a fresh air charge can be drawn in. Hence, part of the return or suction stroke of the piston is non-effective. This non-effective part of the suction stroke must be kept as small as possible in order to keep capacity to a maximum.

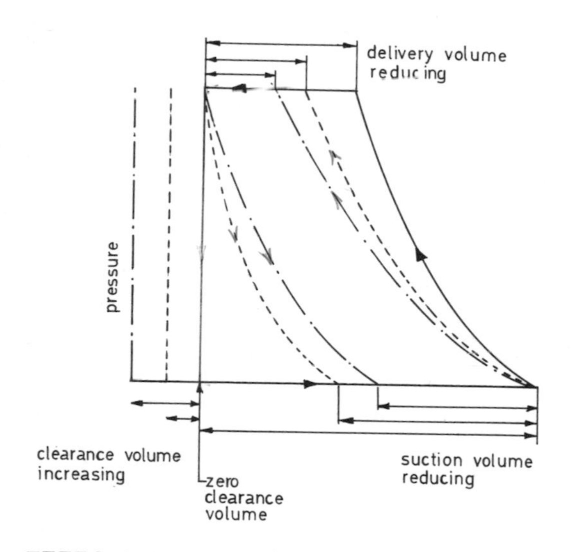

Fig. 86

Volumetric efficiency is a measure of compressor capacity, it is the ratio of the actual volume of air drawn in each suction stroke to the stroke volume. Fig. 86 shows what would happen to the compressor volumetric efficiency — and hence capacity — if the clearance volume were increased.

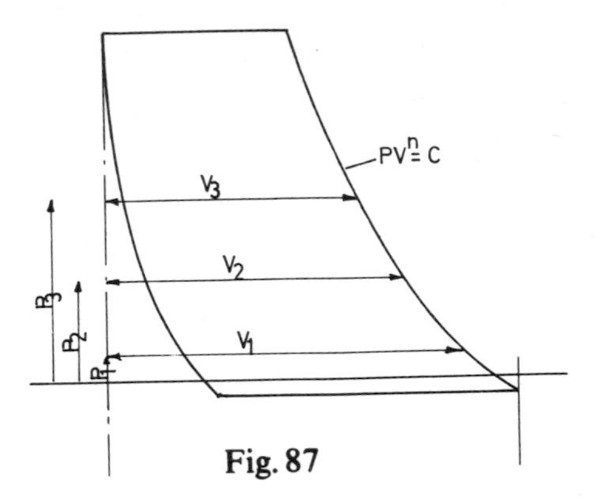

Fig. 87

Clearance volume can be calculated from an indicator card by taking any three points on the compression curve such that their pressures are in geometric progression, *i.e.* $P_1 / P_2 = P_2 / P_3$ hence $P_2 = \sqrt{P_1 P_3}$. (Fig. 87). If V_c = clearance volume as a percentage of the readily calculable stroke volume and V_1, V_2, V_3 are also percentages of the stroke volume then:

$$P_1 (V_1 + V_c)^n = P_2 (V_2 + V_c)^n = P_3 (V_3 + V_c)^n$$

i.e. $$\frac{P_1}{P_2} = \left(\frac{V_2 + V_c}{V_1 + V_c}\right)^n \quad \text{and} \quad \frac{P_2}{P_3} = \left(\frac{V_3 + V_c}{V_2 + V_c}\right)^n$$

now $$\frac{P_1}{P_2} = \frac{P_2}{P_3}$$

therefore $$\left(\frac{V_2 + V_c}{V_1 + V_c}\right)^n = \left(\frac{V_3 + V_c}{V_2 + V_c}\right)^n$$

hence $$\frac{V_2 + V_c}{V_1 + V_c} = \frac{V_3 + V_c}{V_2 + V_c}$$

$$V_c = \frac{V_2^2 - V_1 V_3}{V_1 + V_3 - 2V_2}$$

Since V_1, V_2 and V_3 are known V_c can be calculated.

Correct clearance must be maintained and this is usually done by checking the mechanical clearance and adjusting it as required by using inserts under the palm of the connecting rod. Bearing clearances should also be kept at recommended values.

Methods of ascertaining the mechanical clearance in an air compressor:

1. Remove suction or discharge valve assembly from the unit and place a small loose ball of lead wire on the piston edge, then rotate the flywheel by hand to take the piston over top dead centre. Remove and measure the thickness of the lead wire ball.

2. Put crank on top dead centre, slacken or remove bottom half of the bottom end bearing. Rig a clock gauge with one contact touching some under part of the piston or piston assembly and the other on the crank web. Take a gauge reading. Then by using a suitable lever bump the piston, *i.e.* raise it until it touches the cylinder cover. Take another gauge reading, the difference between the two readings gives the mechanical clearance.

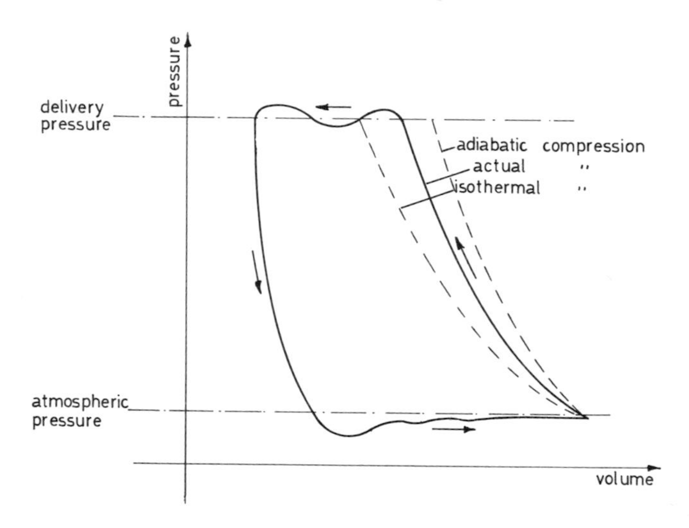

ACTUAL PRESSURE-VOLUME DIAGRAM

Fig. 88

In practice the effective volume drawn in per stroke is further reduced since the pressure in the cylinder on the suction stroke must fall sufficiently below the atmospheric pressure so that the inertia and spring force of the suction valve can be overcome. Fig. 88 shows this effect on the actual indicator card and also the excess pressure above the mean required upon delivery, to overcome delivery valve inertia and spring force.

Air compressors are either reciprocating or rotary types, the former are most commonly used at sea for the production of air for purposes outlined at the beginning of this chapter. The latter types are used to produce large volumes of air at relatively low pressure and are used at sea as integral parts of main engines for scavenging and for boiler forced draught.

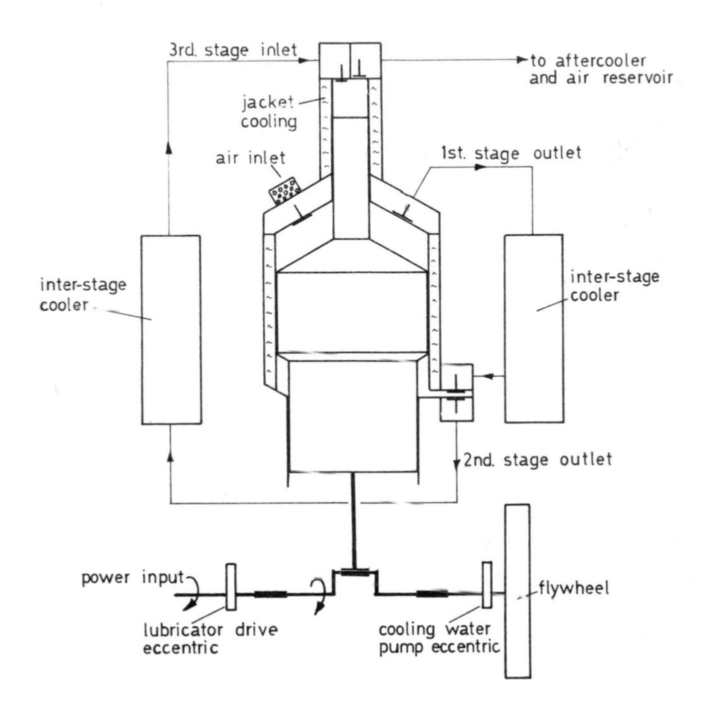

3 STAGE AIR COMPRESSOR

Fig. 89

Reciprocating air compressors at sea are generally two or three stage types with inter-stage cooling. Fig 89 shows diagrammatically a tandem type of three stage compressor, the pressures and temperatures at the various points would be roughly as follows:

	Delivery pressure	*Air temperature*	
		Before the coolers	*After the coolers*
First stage	4 bar	110°C	35°C
Second stage	16 bar	110°C	35°C
Third stage	40 bar	70°C	25°C

The above figures are for a salt water temperature of about 16°C. Final air temperature at exit from the after-cooler is generally at or below atmospheric temperature.

Drains

Fitted after each cooler is a drain valve, these are essential. To emphasise, if we consider 30 m^3 of free air relative humidity 75% temperature 20°C being compressed every minute to about 10 bar, about ½ litre of water would be obtained each minute.

Drains and valves to air storage unit must be open upon starting up the compressor in order to get rid of accumulated moisture. When the compressor is running drains have to be opened and closed at regular intervals.

Filters

Air contains suspended foreign matter, much of which is abrasive. If this is allowed to enter the compressor it will combine with the lubricating oil to form an abrasive-like paste which increases wear on piston rings, liners and valves. It can adhere to the valves and prevent them from closing properly, which in turn can lead to higher discharge temperatures and the formation of what appears to be a carbon deposit on the valves, etc. Strictly, the apparent carbon deposit on valves contains very little carbon from the oil, it is mainly solid matter from the atmosphere.

These carbon like deposits can become extremely hot on valves which are not closing correctly and could act as ignition points for air-oil vapour mixtures, leading to possible fires and explosions in the compressor.

Hence air filters are extremely important, they must be regularly cleaned and where necessary renewed and the compressor must never be run with the air intake filter removed.

Relieving Devices

After each stage of compression a relief valve will normally be fitted. Regulations only require the fitting of a relieving device on the h.p. stage. Bursting discs or some other relieving device are fitted to the water side of coolers so that in the event of a compressed air carrying tube bursting, the sudden rise in pressure of the surrounding water will not fracture the cooler casing. In the event of a failure of a bursting disc a thicker one must not be used as a replacement.

Lubrication

Certain factors govern the choice of lubricant for the cylinders of an air compressor, these are:

Operating temperature, cylinder pressures and air condition.

Operating temperature effects oil viscosity and deposit formation. If the temperature is high this results in low oil viscosity, very easy oil distribution, low film strength, poor sealing and increased wear. If the temperature is low, oil viscosity would be high, this causes poor distribution, increased fluid friction and power loss.

Cylinder pressures. If these are high the oil requires to have a high film strength to ensure the maintenance of an adequate oil film between the piston rings and the cylinder walls.

Air condition. Air contains moisture that can condense out. Straight mineral oils would be washed off surfaces by the moisture and this could lead to excessive wear and possible rusting. To prevent this a compounded oil with a rust inhibitor additive would be used. Compounding agents may be from 5 to 25% of non-mineral oil, which is added to a mineral oil blend. Fatty oils are commonly added to lubricating oil that must lubricate in the presence of water, they form an emulsion which adheres to the surface to be lubricated.

Two Stage Air Compressor

Most modern diesel engines use starting air at a pressure of about 26 bar and to achieve this a two stage type of compressor would be adequate. These compressors are generally of the reciprocating type, with various possible arrangements of the cylinders, or they could be a combination of a rotary first stage followed by a reciprocating high pressure stage. This latter arrangement leads to a compact, high delivery rate compressor.

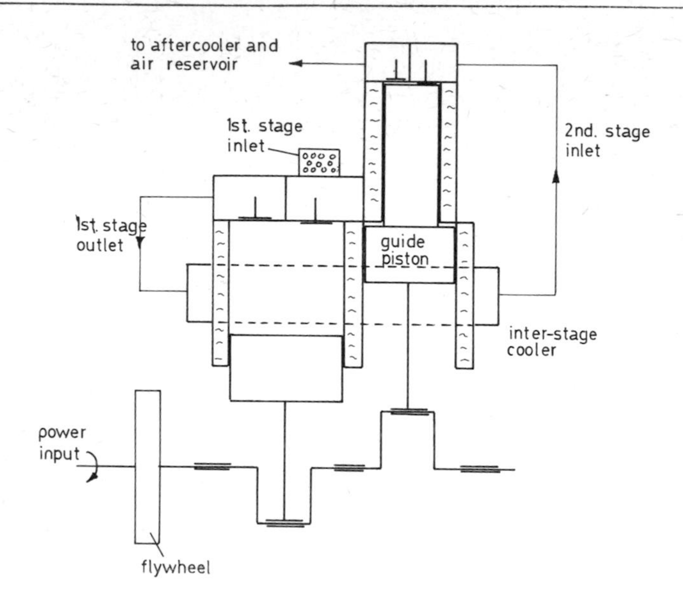

2 STAGE AIR COMPRESSOR

Fig. 90

Fig 90 shows a typical two-stage reciprocating type of air compressor, the pressures and temperatures at the various points would be approximately as follows:

	Delivery pressure	*Air temperature Before the coolers*	*After the coolers*
First stage	$4\frac{2}{3}$ bar	130°C	35°C
Second stage	$26\frac{2}{3}$ bar	130°C	35°C

Compressor Valves

Simple suction and discharge valves are shown in Fig. 91, these would be suitable diagrams for reproduction in an examination. Modern valves are somewhat more streamlined and lighter in order to reduce friction losses and valve inertia. Materials used in the construction are generally:

Valve seat. 0.4% carbon steel hardened and polished working surfaces.

Valve. Nickel steel, chrome vanadium steel or stainless steel, hardened and ground, then finally polished to a mirror finish.

Spring, hardened steel. (N.B. all hardened steel would be tempered).

Valve leakages do occur in practice and this leads to loss of efficiency and increase in running time.

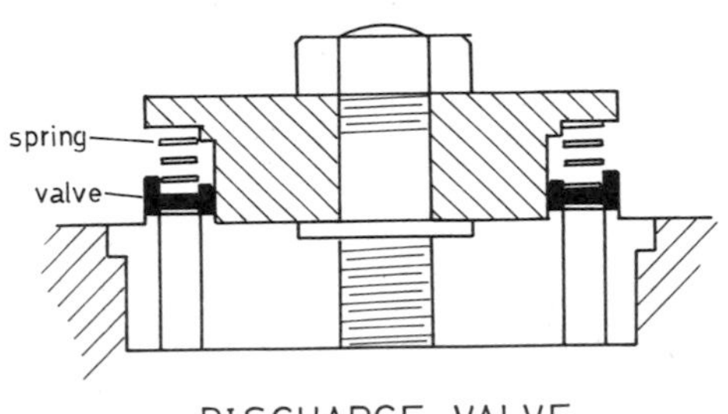

DISCHARGE VALVE

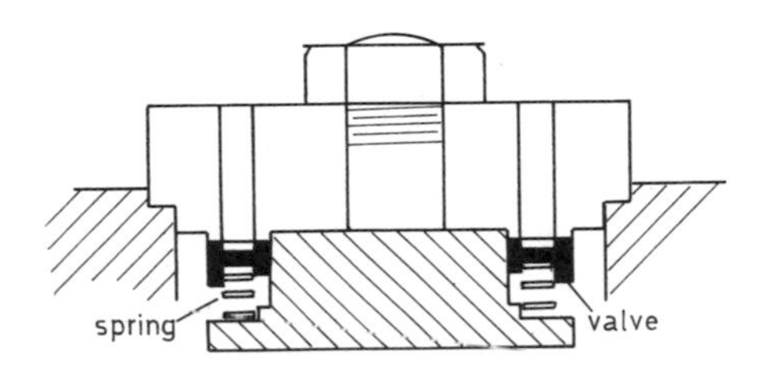

SUCTION VALVE

COMPRESSOR VALVES

Fig. 91

Effects of Leaking Valves

1. *First stage suction*

Reduced air delivery, increased running time and reduced pressure in the suction to the second stage. If the suction valve leaks badly it may completely unload the compressor.

2. *First stage delivery*

With high pressure air leaking back into the cylinder less air can be drawn in, this means reduced delivery and increased discharge temperature.

3. *Second stage suction*

High pressure and temperature in the second stage suction line, reduced delivery and increased running time.

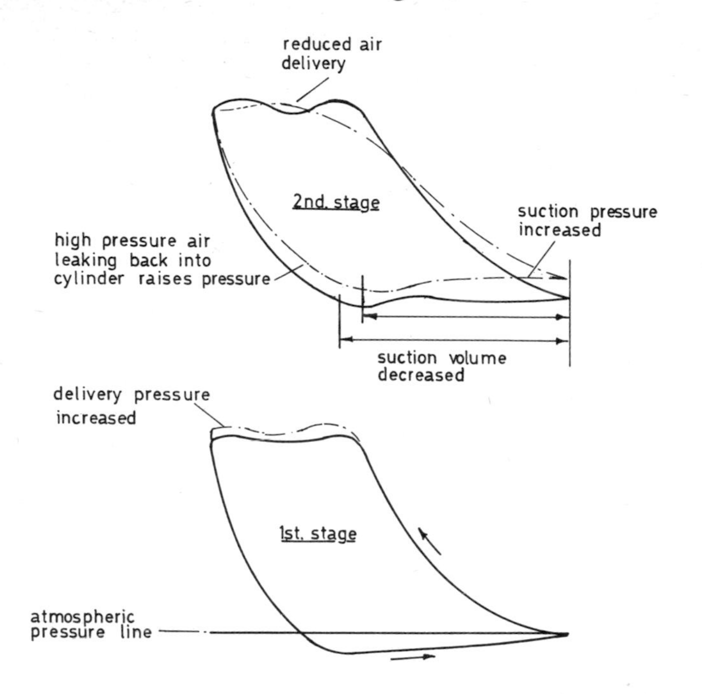

EFFECT OF LEAKING 2nd. STAGE DELIVERY VALVES

Fig. 92

4. *Second stage delivery*

Increased suction pressure in second stage, reduced air suction and delivery in second stage. Delivery pressure from first stage increased. Fig. 92 shows the effect of a leaking second stage delivery valve on the indicator cards of a compressor.

It must be remembered that it is not usual to find a facility for taking indicator cards from air compressors.

Regulation of Air Compressors

Various methods are available:

Start stop control. This is only suitable for small electrically driven types of unit. A pressure transducer attached to the air receiver set for desired max-min pressures would switch the current to the electric motor either on or off. Drainage would have to be automatic and air receiver relatively large compared to the compressor unit requirements so that the number of starts per unit time is not too great. It must be remembered that the starting current for an electric motor is about double the normal running current.

Constant running control. This method of control is the one most often used. The compressor runs continuously at a constant speed and when the desired air pressure is reached the air compressor is unloaded in some way so that no air is delivered and practically no work is done in the compressor cylinders.

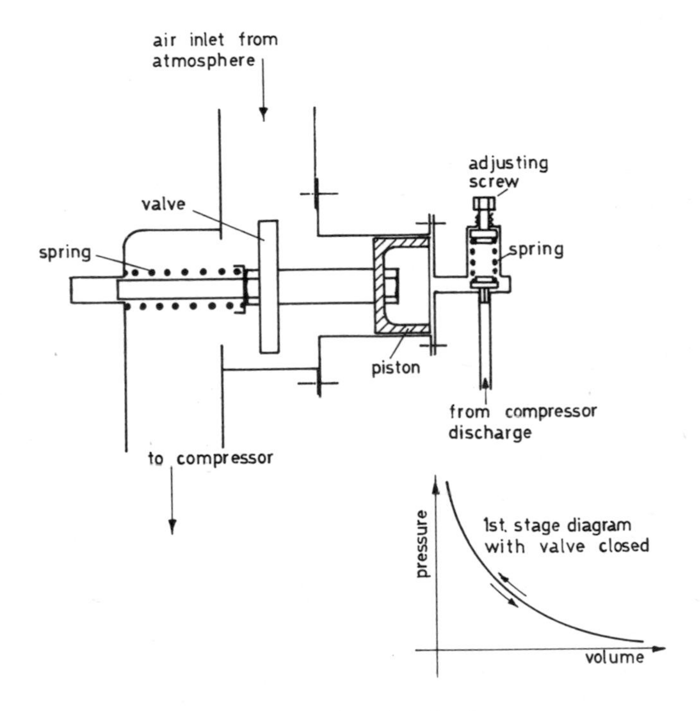

COMPRESSOR UNLOADING VALVE

Fig. 93

The methods used for compressor unloading vary, but that most commonly used is on the suction side of the compressor. If the compressor receives no air then it cannot deliver any. Or if the air taken in at the suction is returned to the suction no air will be delivered. In either case virtually no work would be done in the compressor cylinder or cylinders and this would provide an economy compared to discharging high pressure air to the atmosphere through a relief valve.

Fig. 93 shows diagrammatically a compressor unloading valve fitted to the compressor suction. When the discharge air pressure reaches a desired value it will act on the piston causing the spring loaded valve to close shutting off the supply of air to the compressor. An alternative arrangement to the foregoing is to keep the compressor non-return spring loaded suction valves closed or open to unload the compressor.

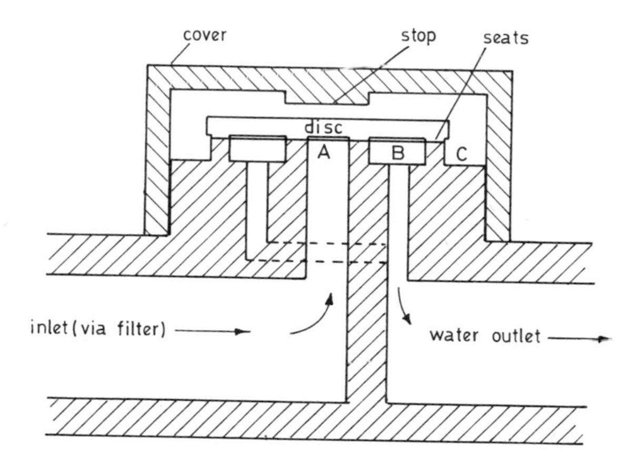

AIR DRAIN TRAP

Fig. 94

Automatic Drain

Fig. 94 shows an automatic air drain trap which functions in a near similar way to a steam trap.

With water under pressure at the inlet the disc will lift, allowing the water to flow radially across the disc from A to the outlet B. when the water is discharged and air now flows radially outwards from A across the disc, the air expands increasing in velocity ramming air into C and the space above the disc, causing the disc to close on the inlet. Because of the build-up of static pressure in the space above the disc in this way, and the differential area on which the pressures are acting, the disc is held firmly closed. It will remain so unless the pressure in the space above the disc falls.

In order that this pressure can fall, and the trap re-open, a small groove is cut across the face of the disc communicating B and C through which the air slowly leaks to outlet.

Obviously this gives an operational frequency to the opening and closing of the disc which is a function of various factors, *e.g.* size of groove, disc thickness, volume of space above the disc, etc. It is therefore essential that the correct trap be fitted to the drainage system to ensure efficient and effective operation.

AIR VESSELS

Material used in the construction must be of good quality low carbon steel similar to that used for boilers, *e.g.* 0.2% Carbon (max.), 0.35% Silicon (max.), 0.1% Manganese, 0.05% Sulphur (max.), 0.05% Phosphorus (max.), u.t.s. 460 MN/m^2.

Welded construction has superseded the riveted types and welding must be done to class 1 or class 2 depending upon operating pressure. If above 35 bar approximately then class 1 welding regulations apply.

Some of the main points relating to class 1 welding are that the welding must be radiographed, annealing must be carried out at a temperature of about 600°C and a test piece must be provided for bend, impact and tensile tests together with micrographic and macrographic examination.

Mountings generally provided are shown in Fig. 95. If it is possible for the receiver to be isolated from the safety valve then it must have a fusible plug fitted, melting point approximately 150°C, and if carbon dioxide is used for fire fighting it is recommended that the discharge from the fusible plug be led to the deck. Stop valves on the receiver generally permit of slow opening to avoid rapid pressure increases in the piping system, and piping for starting air has to be protected against the possible effects of

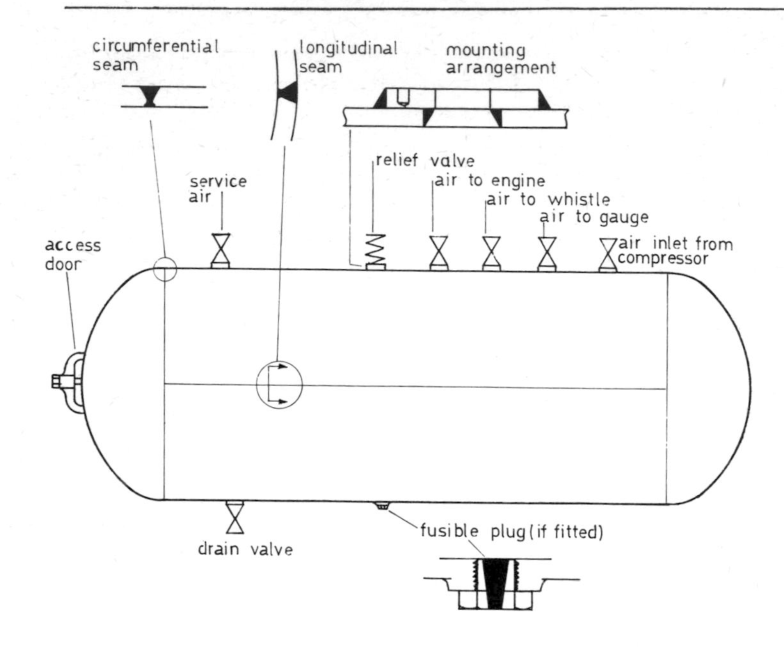

Fig. 95

explosion. Drains for the removal of accumulated oil and water are fitted to the compressor, filters, separators, receivers and lower parts of pipe-lines.

Before commencing to fill the air vessel after overhaul or examination, ensure:

1. Nothing has been left inside the air vessel, *e.g.* cotton waste that could foul up drains or other outlets.
2. Check pressure gauge against a master gauge.
3. All doors are correctly centred on their joints.

Run the compressor with all drains open to clear the lines of any oil or water, and when filling open drains at regular intervals, observe pressure.

After filling close the air inlet to the bottle, check for leaks and follow up on the door joints.

When emptying the receiver prior to overhaul, etc., ensure that it is isolated from any other interconnected receiver which must, of course, be in a fully charged state.

Cleaning the air receiver internally must be done with caution. Any cleaner which gives off toxic, inflammable or obnoxious fumes should be avoided. A brush down and a coating on the internal surfaces of some protective, harmless to personnel, such as a graphite suspension in water could be used.

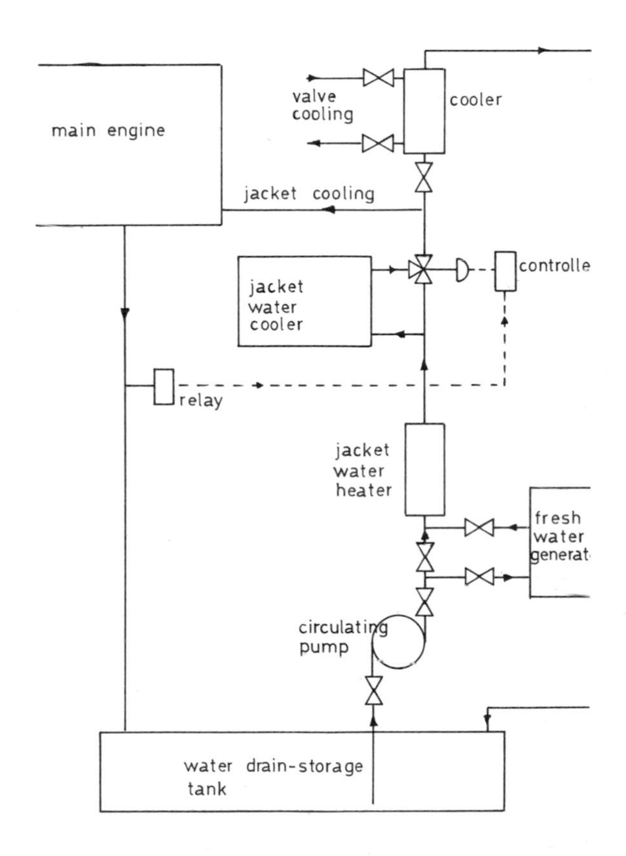

Fig. 96

COOLING SYSTEMS

These can conveniently be grouped into sections.

1. *Cylinder cooling,* or jacket cooling: normally fresh or distilled water. This may incorporate cooling of the turbine or turbines in a turbocharged engine and exhaust valve cooling.

2. *Fuel valve cooling*: this would be a separate system using fresh water or a fine mineral oil.

3. *Piston cooling*: this may be lubricating oil, distilled or fresh water. If it is oil the system is generally common with the lubrication system. If water, a common storage tank with the jacket cooling system would generally be used.

4. *Charge air cooling*: this is normally sea water.

Comparison of coolants

1. *Fresh water*: Inexpensive, high specific heat, low viscosity. Contains salts which can deposit, obstruct flow and cause corrosion. Requires treatment. Leakages could contaminate lubricating oil system leading to loss of lubrication, possible overheating of bearings and bearing corrosion. Requires a separate pumping system.

It is important that the water should not be changed very often as this can lead to increased deposits. Leakages from the system must be kept to an absolute minimum, so a regular check on the replenishing-expansion tank contents level is necessary.

If the engine has to stand inoperative for a long period and there is a danger of frost, (a) drain the coolant out of the system, (b) heat up the engine room, or (c) circulate system with heating on.

It may become necessary to remove scale from the cooling spaces, the following method could be used. Circulate, with a pump, a dilute hydrochloric acid solution. A hose should be attached to the cooling water outlet pipe to remove gases. Gas emission can be checked by immersing the open end of the hose occasionally into a bucket of water. Keep compartment well ventilated as the gases given off can be dangerous. Acid solution strength in the system can be tested from time to time by putting some on to a piece of lime. When the acid solution still has some strength and no more gas is being given off then the system is scale free. The system should now be drained and flushed out with fresh water, then neutralised with a soda solution and pressure tested to see that the seals do not leak.

2. *Distilled water:* More expensive than fresh water, high specific heat, low viscosity. If produced from evaporated salt water it would be acidic. No scale forming salts. Requires separate pumping system. Leakages could contaminate the lubricating oil system, causing loss of lubrication and possible overheating and failure of bearings, etc.

Additives for cooling water

Those generally used are either anti-corrosion oils or inorganic inhibitors.

If pistons are water cooled an anti-corrosion oil is recommended as it lubricates parts which have sliding contact. The oil forms an emulsion and part of the oil builds up a thin unbroken film on metal surfaces, this prevents corrosion but is not thick enough to impair heat transfer.

Inorganic inhibitors form protective layers on metal surfaces guarding them against corrosion.

It is important that the additives used are not harmful if they find their way into drinking water — this is possible if the jacket cooling water is used as a heating medium in a fresh water generator. Emulsion oils and sodium nitrite are both approved additives, but the latter cannot be used if any pipes are galvanised or if any soldered joints exist. Chromates cannot be used if the cooling water is used in a fresh water generator and it is a chemical that must be handled with care.

3. *Lubricating oil:* Expensive. Generally no separate pumping system required since the same oil is normally used for lubrication and cooling. Leakages from cooling system to lubrication system are relatively unimportant providing they are not too large, otherwise one piston may be partly deprived of coolant with subsequent overheating.

Due to reciprocating action of pistons some relative motion between parts in contact in the coolant supply and return system must occur, oil will lubricate these parts more effectively than water. No chemical treatment required. Lower specific heat than water, hence a greater quantity of oil must be circulated per unit time to give the same cooling effect.

If lubricating oil encounters high temperature it can burn, leaving as it does so carbon deposit. This deposit on the underside of a piston crown could lead to impairment of heat transfer, overheating and failure of the metal. Generally the only effective method of dealing with the carbon deposit is to dismantle the piston and physically remove it. Since oil can burn in this way a lower mean outlet and inlet temperature for the oil has to be maintained, in order to achieve this more oil must be circulated per unit time.

Some engines may use completely separate systems for oil cooling of pistons and bearing lubrication, the advantages gained by this method are:

1. Different oils can be used for lubrication and cooling, a very low viscosity mineral oil would be better suited to cooling than lubrication.

2. Additives can be used in the lubricating oil that would be beneficial to lubrication, *e.g.* oiliness agents, e.p. agents and V.I. improvers, etc.

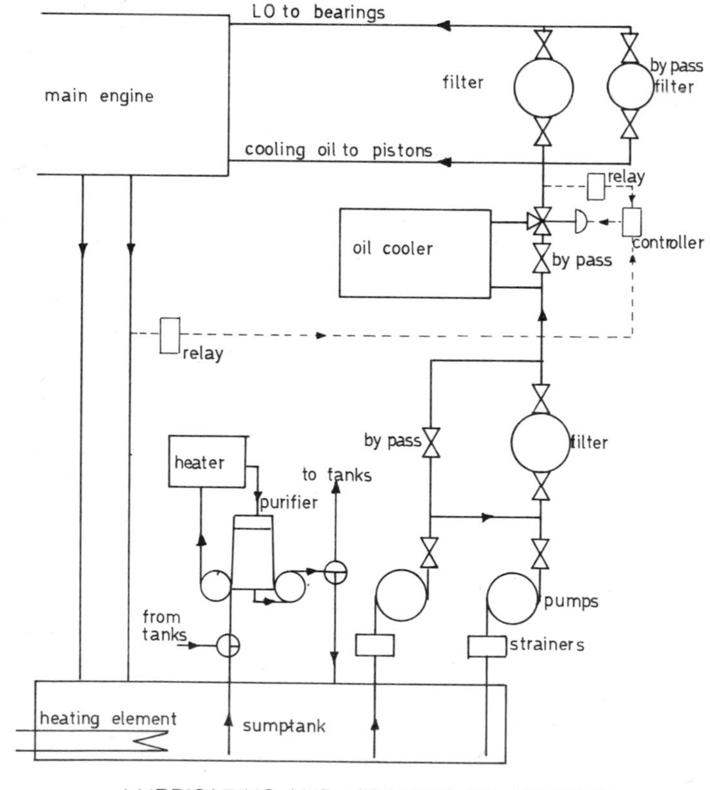

LUBRICATING AND COOLING OIL SYSTEM

Fig. 97

3. Improved control over piston temperatures.

4. If oil loss occurs, then with separate systems the problem of detection is simplified and in the case of total oil loss in either system, the quantity to be replaced would not be as great as for a common system.

5. Contamination of the oil in either system may take place. In the event the problem of cleaning or renewal of the oil is not so great.

6. Oxidation of lubricating oil in contact with hot piston surfaces leads to rapid reduction in lubrication properties.

Disadvantages of having two separate systems are: Greater initial cost due to separate storage, additional pipework and pumps. A sealing problem to prevent mixing of the two different oils is created and due to the increased complexity more maintenance would have to be carried out.

TEST EXAMPLES

Second Class

1. Enumerate all the fittings provided on an air receiver, explain for what purpose they are provided and how they are fitted. What precautions would you take when, (a) filling, (b) emptying, (c) opening up, (d) cleaning the air receiver.

2. Sketch and describe a two stage air compressor. Give reasons for fitting the following: suction filter, bursting discs, drains. What would be the effects of excessive clearance volume in the two stages?

3. An air compressor has to have a new bottom end bearing fitted. Describe the procedure for the fitting and the precautions to be taken before and after starting the compressor.

4. Sketch and describe a suction and a delivery valve for an air compressor. Show clearly how they are fitted in the cylinder head and describe the overhaul of one of the valves.

5. With reference to crankcase lubricating oil discuss the following: (a) acidity, (b) dilution by fuel oil, (c) water leakage. Describe any tests that could be made to detect the foregoing contamination.

6. Sketch and describe a jacket and piston cooling water system for a marine diesel engine. How is expansion of the water allowed for?

First Class

1. Sketch and describe a two stage reciprocating air compressor, insert on your diagram pressures and temperatures at the salient points. Describe the effects of leaking second stage discharge valves and alteration in clearance volume.

2. Sketch and describe a three stage reciprocating air compressor giving temperatures and pressures at the salient points in the cycle. Why is air cooling necessary? Explain how clearances may be obtained and adjusted. Describe any safety features which guard against excess pressures.

3. Make a diagrammatic sketch of a three stage air compressor. Explain with the aid of a P-V diagram the effects of inter-stage cooling. How would the following effect the running of the compressor?
(a) excessive cylinder clearance, (b) choked air intake filter, (c) leaking high pressure stage suction valve, (d) leaking low pressure inlet valve, (e) water precipitation.

4. Make a linc diagram of an air receiver for a large marine diesel engine. Enumerate all the fittings provided explaining their purpose. What could be the effects of insufficient drainage?

5. Compare fresh water, distilled water and lubricating oil as cooling media for marine diesel engines. What is the cause of carbon deposits, how are they prevented and removed? What treatment is used for distilled water? Give reasons.

6. Enumerate the advantages and disadvantages of having a separate system for oil cooled pistons and bearing lubrication.

CHAPTER 8

MEDIUM SPEED DIESELS

The term medium speed refers to diesels that operate within the approximate speed range 300 to 800 revolutions per minute. High speed is usually 1000 rev/min and above.

A considerable swing towards this type of engine has been taking place over the last decade and at present many prototypes of increased capability are operating or are under construction that when finally marketed could greatly accelerate the swing. If the advantages and salient features of the medium speed diesel are examined the reader will appreciate why this swing, for certain vessel tonnages, is taking place. They are as follows:

1. Compact and space saving. Vessel can have reduced height and broader beam — useful in some ports where shallow draught is of importance. The considerable reduction in engine height compared to direct drive engines and the reduced weight of components means that lifting tackle, such as the engine room crane, is reduced in size as it will have lighter loads to lift through smaller distances. More cargo space is made available and because of the higher power weight ratio of the engine a greater weight of cargo can be carried.

2. Through using a reduction gear a useful marriage between ideal engine speed and ideal propeller speed can be achieved. For optimum propeller speed hull form and rudder have to be considered, the result is usually a slow turning propeller (for large vessels this can be as low as 50 to 60 rev/min). Gearing enables the Naval Architect to design the best possible propeller for the vessel without having to consider any dictates of the engine.

Engine designers can ignore completely propeller speed and concentrate solely upon producing an engine that will give the best possible power weight ratio.

3. Modern tendency is to utilise uni-directional medium speed geared diesels coupled to either a reverse reduction gear, controllable pitch propeller or electric generator. The first two of these

methods are the ones primarily used and the advantages to be gained are considerable, they are:

(a) Less starting torque required, clutch disengaged or controllable pitch propeller in neutral.

(b) Reduced number of engine starts, hence starting air capacity can be greatly reduced and compressor running time minimised. Classification society requirements are six consecutive starts without air replenishment for non-reversible engines and twelve for reversible engines. Cylinder liner wear will also be reduced, it is well known that the greatest cylinder liner wear rate occurs upon starting.

(c) Engines can be tested at full speed with the vessel alongside a quay without having to take any special precautions.

(d) With the engine or engines running continuously, power can be taken off via a clutch or clutch/gear drive for the operation of electric generators or cargo pumps, etc. Hence the engine has become a multi-purpose 'power pack.'

(e) Improved manoeuvrability, vessel can be brought to rest within a shorter distance by intelligent use of the engines and c.p. propeller.

(f) Staff load during 'stand-by' periods is reduced, system lends itself ideally to simple bridge control.

4. With two engines coupled via gearing one may be disengaged, whilst the other supplies the motive power, and overhauled. This reduces off hire time and voyage is continued at slightly reduced speed with a fuel saving.

5. Spare parts are easier to store and manhandle, unit overhaul time will be greatly reduced.

ENGINE COUPLINGS, CLUTCHES AND GEARING

Various arrangements of geared coupled engines are possible, the basic arrangement depends upon the services the engine has to supply, *e.g.* a high electrical load in port may have to be catered for with the alternator being driven at a higher speed than the engine. Hence a step up gear box would be required along with some form of clutch. Large capacity cargo pumps operating at high speed would require a similar arrangement. Fig. 98 shows different types of arrangements with different types of clutches or couplings being used.

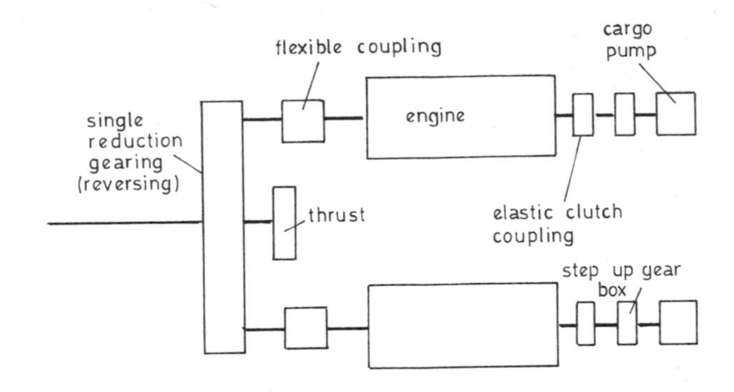

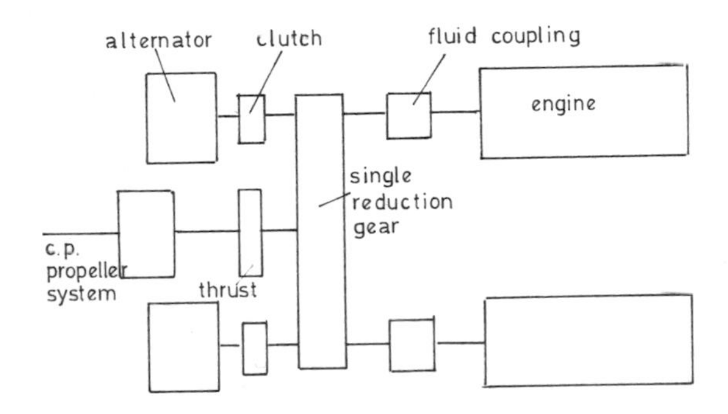

ENGINE ARRANGEMENTS

Fig. 98

Fluid Couplings

These are completely self-contained, apart from a cooling water supply, they require no external auxiliary pump or oil feed tank. A scoop tube when lowered picks up oil from the rotating casing reservoir and supplies it to the vanes for coupling and power transmission, withdrawal of the scoop tube from the oil stops the flow of oil to the vanes which then drains to the reservoir. During power transmission a flow of oil takes place continuously through the cooler and clutch.

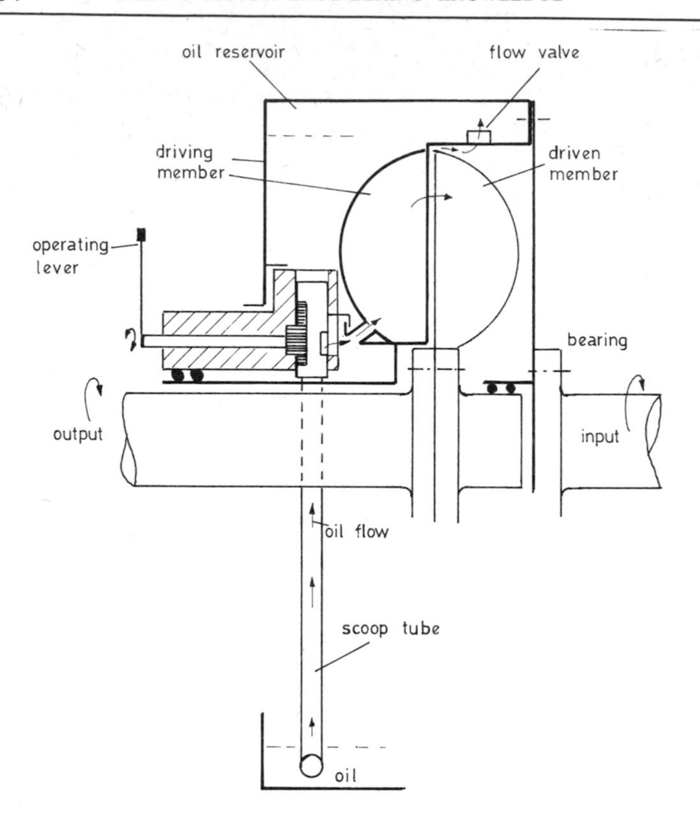

Fig. 99

Fluid clutches operate smoothly and effectively. They use a fine mineral lubricating oil and have no contact and hence no wear between driving and driven members. Torsional vibrations are dampened out to some extent by the clutch and transmitted speeds can be considerably less than engine speed if required, by suitable adjustment of the scoop tube. It is possible to have a dual entry scoop tube for reversible engines, this obviates the use of c.p. propellers or reversible reduction gears but the control problem is considerably more complex with reversible engines, they have to be stopped and started and if 4-stroke engines are used camshafts have to be moved, etc.

Reverse Reduction Gear

These gear systems are mainly restricted, at present, to powers of up to about 4800kW for twin engined single screw installations. There obvious advantages are:

1. Uni-directional engine.
2. No c.p. propeller required.
3. Ability to engage or disengage either engine of a twin engine installation from the bridge by a relatively simple remote control.
4. Improved manoeuvrability, etc.

When dealing with large powers the friction clutches used in the system can become excessively large, great heat generation during engagement may require a cooling system, the whole becomes more expensive and it may be cheaper to use direct reversing engines — however it would for reasons previously outlined be more prudent to use a c.p. propeller.

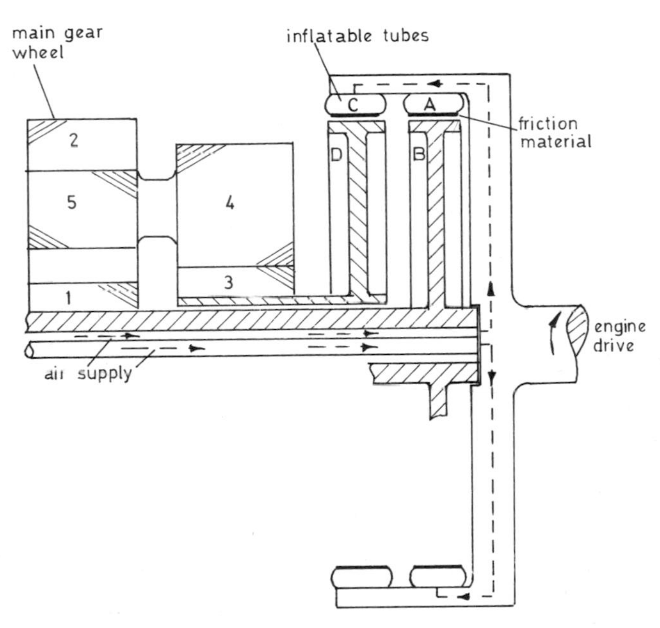

FRICTION CLUTCH WITH SINGLE REDUCTION REVERSING GEAR

Fig. 100

Two systems of reverse reduction gear are shown in Figs 100 and 101. In Fig 100 the engine drives a steel drum which has two inflatable synthetic rubber tubes bonded to its inner surface. These tubes have friction material, like brake lining, on their inner surface. Air is supplied through the centrally arranged tube, or the annulus formed by the tube and shaft hole to one or the other of the inflatable tubes. Two flanged wheels are connected via hollow shafts and gears to the main gear wheel and shaft.

For operation ahead, air would be supplied to inflatable tube A which would then by friction on flanged wheel B bring gears 1 and 2 up to speed, gears 3, 4 and 5 together with flanged wheel D would be idling.

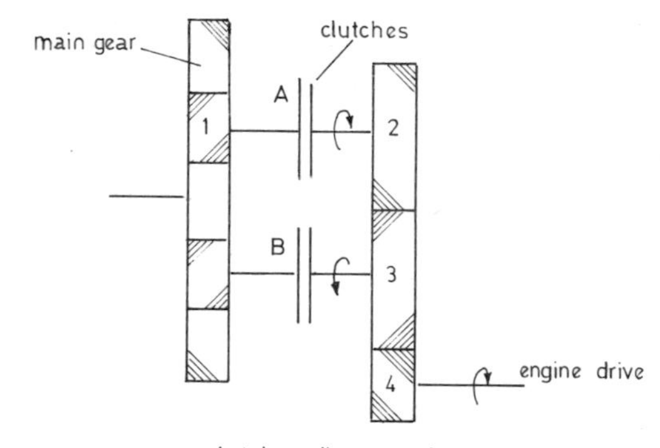

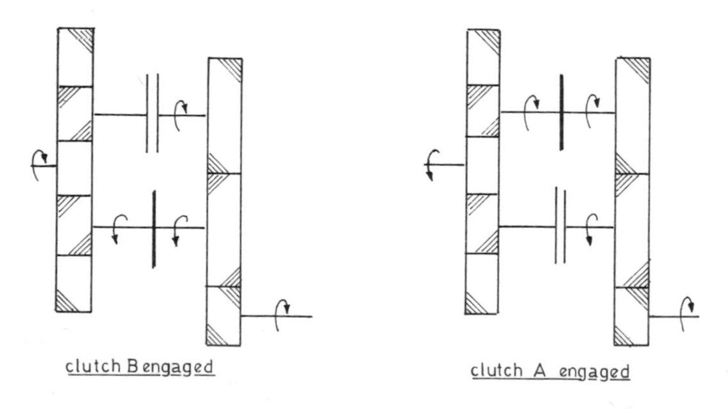

REVERSIBLE REDUCTION GEAR

Fig. 101

For astern operation, air would be supplied to inflatable tube C (A evacuated) and by friction on flanged wheel D gears 3, 4, 5 and 2 would be brought up to speed, gear 1 and drum B would be idling. For single reduction, gears 3 and 4 would be the same size and so would be gears 1 and 5.

An alternative system, either single or double reduction but probably the latter, is shown in Fig. 101. Friction clutches A and B are pneumatically controlled from some remote position. Gears 1, 2, 3 and 4 would have to be the same size if the gear were to be single reduction — but this is most unlikely.

Flexible Couplings

These are used between engine and gearbox to dampen down torque fluctuations, reduce the effects of shock loading on the gears and engine, cater for slight misalignments. They are also used in conjunction with clutches for power take-off when required. In construction they may be similar to the well known

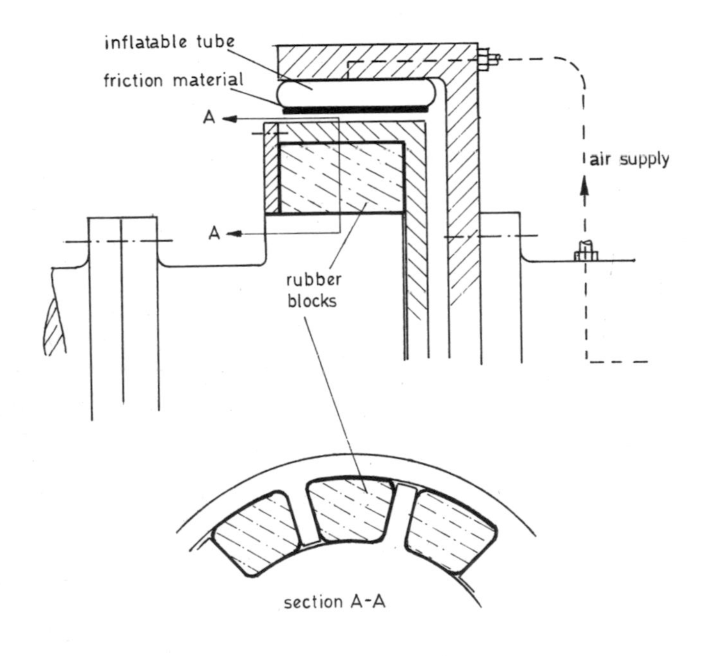

FLEXIBLE CLUTCH COUPLING

Fig. 102

multi-tooth type to be found in turbine installations or employ diaphragms or rubber blocks. Those types that use rubber or synthetic rubber, such as Nitrile, give electrical insulation between driving and driven members, but all types will minimise vibration and reduce noise level.

Fig. 102 shows a combination of flexible coupling and pneumatically operated friction clutch, the arrangement gives a smooth transition of speed and torque during engagement, it could be typical of an arrangement for take off for electrical power or cargo pumps, etc. The rubber blocks would be synthetic if oil is likely to be present as natural rubber is attacked by oil.

EXHAUST VALVES

Most 4-stroke medium speed diesels incorporate two exhaust valves per cylinder and if we consider a moderate sized installation consisting of two 12-cylinder V engines this gives a total of 48 exhaust valves. A not inconsiderable quantity, and if the plant is to burn fuel of high viscosity, the maintenance problem for these valves could be considerable.

In order to minimise maintenance and to prolong valve life, bearing in mind that burning of high viscosity oil is essential due to the soaring cost of light diesel oil, certain design parameters and operating procedures must be followed. These are:

1. Separately caged exhaust valves are preferred even though they increase first cost. If they are made integral with the cylinder head and a fuel of poorer quality than normal has to be burnt, increased frequency of replacement and overhaul of the valves necessitating cylinder head removal each time becomes a tedious time taking operation.

2. All connections to the valves, cooling, exhaust, etc. should be capable of easy disconnection and re-assembly. Use of combined cooling inlet and return injection point could prove useful.

3. Materials that have to operate at elevated temperatures must be capable of withstanding the erosive and corrosive effects of the exhaust gas. When burning oils of high viscosity which contain Sodium and Vanadium, deposits can form on the valve seats which at high temperatures (in excess of 530°C at the valve seat) become strongly corrosive sticky compounds which lead to burnt valves. Hence the need for materials that can withstand the corrosion and for intense cooling arrangements for valve seats.

Stellited valve seats are not uncommon, Stellite is a mixture of Cobalt, Chromium and Tungsten extremely hard and corrosion resistant that is fused on to the operating surfaces.

Low temperature corrosion due to sulphur compounds can occur during prolonged periods of running under low load conditions. The valve spindle and guide, which would be at a relatively low temperature, are the principal places of attack due to the effective cooling in this region. Ideally, valve cooling should be a function of engine load with the valve being maintained at a uniform temperature at all times, this could prove complicated and expensive to arrange for.

4. Effective lubrication of the valve spindle is necessary to avoid risk of seizure and possible mechanical damage due to a valve 'hanging up.' In order to minimise lubricating oil usage the lubrication system for the valves would be similar to that used for cylinder lubrication and since the amount of oil used would there-

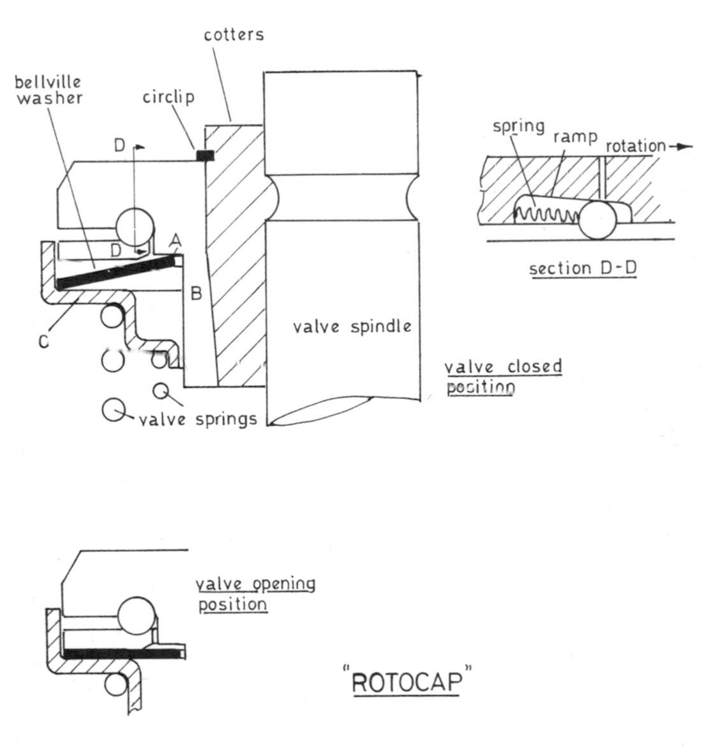

Fig. 103

fore be in small quantities any contamination of the oil by combustion products and water, etc. would be minimal, this would also increase the life of crankcase lubricating oil.

Rotocap

This simple device when fitted to exhaust valves causes rotation of the valve spindle during valve opening, wear of the valve seat is reduced, seat deposits are loosened, valve operational life is

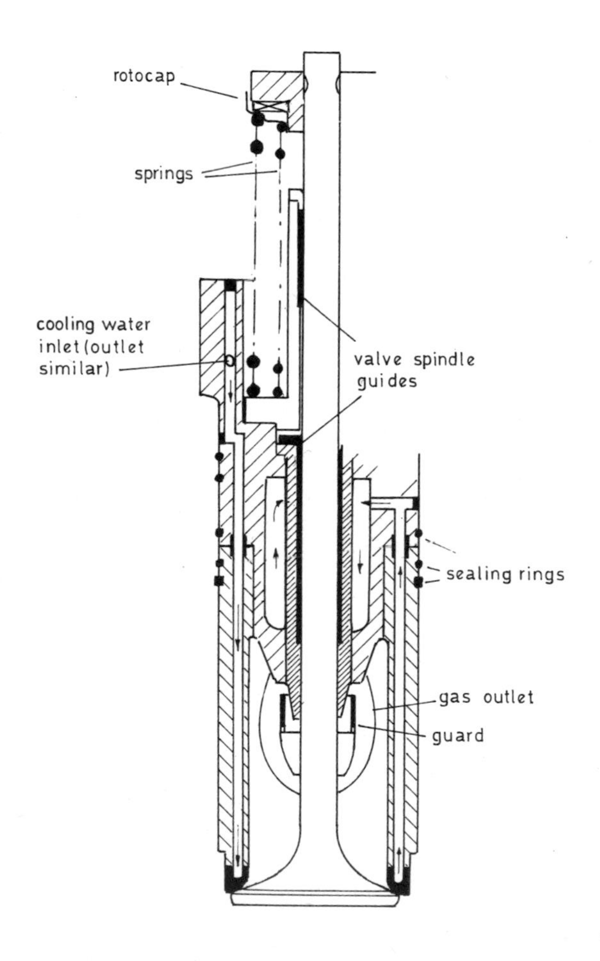

EXHAUST VALVE

Fig. 104

extended. Fig. 103 shows the Rotocap which operates as follows: An increase in spring force on the valve as it opens flattens the belleville washer so that it no longer bears on the bearing housing (B) at A, this removes the frictional holding force between B and C, the spring cover. Further increase in spring force causes the balls to move down the ramps in the retainer imparting as they move a torque which rotates the valve spindle. As the valve closes, the load from the belleville washer is removed from the balls and they return to the position shown in section D-D.

Fig. 104 shows a exhaust valve with welded stellited seat around which cooling water flows keeping the metal temperature at full load conditions well below 500°C, minimising the risk of attack by sodium-vanadium compounds. A guard for the valve spindle

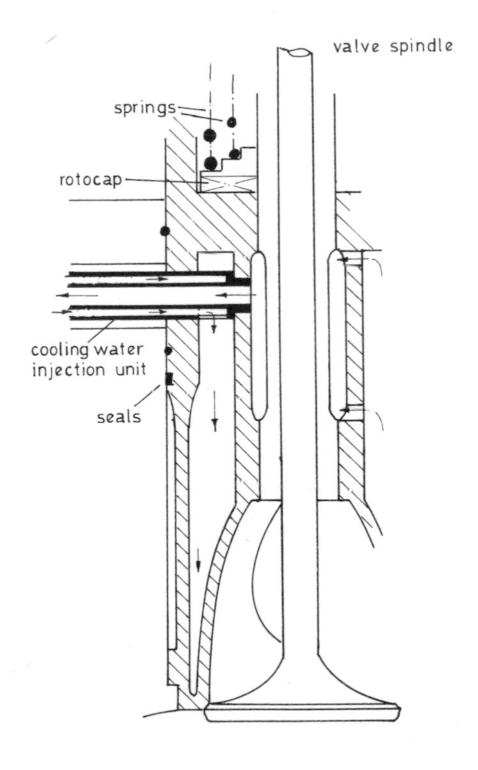

Fig. 105

and guide protects them from direct contact with the hot exhaust gases, this assists cooling and lubrication and reduces risk of seizure.

Fig. 105 shows a different type of exhaust valve but again with intense seat cooling and a rotocap (note its position in the previous figure). By using a combined cooling water inlet and return injection unit the valve can be quickly removed and replaced.

ENGINE DESIGN

The principal design parameters for a medium speed diesel engine are:

1. High power/weight ratio.
2. Simple, strong, compact and space saving.
3. High reliability.
4. Able to burn a wide range of fuels.
5. Easy to maintain, the fact that components are smaller and lighter than those for slow speed diesels makes for easier handling, but accessibility and simple to understand arrangements are inherent features of good design.
6. Easily capable of adaption to unmanned operation.
7. Low fuel and lubricating oil consumption.
8. High thermal efficiency.
9. Low cost and simple to install.

Types of Engine

Either 2- or 4-stroke cycle single acting turbocharged with 'in line' or 'V' cylinder configuration. The main choice is, certainly at present, for the 4-stroke engine and there are various reasons for this.

1. They are capable of operating satisfactorily on the same heavy oils as slow speed 2-stroke engines.

2. Effective scavenging, this is easy to achieve in slow speed 2-stroke engines but it becomes a problem with increase in mean piston speed.

3. Higher mean piston speed. Mean piston speed is simply twice the stroke times the revs/second. For medium speed diesels it

would be approximately 9 to 10 m/s and for slow running diesels 7 to 9 m/s would be about average.

In order that greater power can be developed in the cylinder the working fluid must be passed through faster, hence the higher the mean piston speed for a given unit the greater the power. However, practical limitations govern the piston speed. The relation between cylinder cross sectional area and areas of exhaust and air inlet, method of turbocharging and inertia forces are the main limitations. To reduce inertia forces use is made of aluminium alloy skirted pistons or complete aluminium alloy pistons. Inertia forces must be taken into account for bearing loads — important in trunk piston engines (*i.e.* the majority of medium and high speed diesels) where the guide surface is the cylinder liner, the smaller the side thrust the less the friction and wear.

4. Engine can operate effectively with the turbo-charger out of commission, this would present a considerable problem with some 2-stroke engines of the medium speed type.

5. Turbo-charger size and power can be reduced.

6. It is also claimed that the fuel consumption would be reduced.

Typical 'V' Type Engine

The following is a brief description of a medium speed diesel engine currently in use.

Cylinder bore: 410 mm.

Stroke: 470 mm.

4-stroke cycle supercharged with up to 20 cylinders developing about 500 kW per cylinder at around 600 rev/min.

Bedplate and cylinder blocks are of heavy section cast iron, this gives a strong compact arrangement with good properties for damping out vibrations. As a result of this only two rows of tie bolts are required.

Pistons are aluminium alloy with a cast in cooling coil and piston ring carrier.

Cylinder head incorporates two exhaust valves, water cooled, in removable cages and two air inlet valves with hardened seat inserts. Crankshaft is a solid forging of low alloy-carbon steel, this also applies to the connecting rod.

Liners are of pearlitic iron, simple one piece castings with a deep upper flange drilled for effective cooling. Quills are fitted for separate lubrication.

Bottom end bearings are in three parts, this simplifies removal.

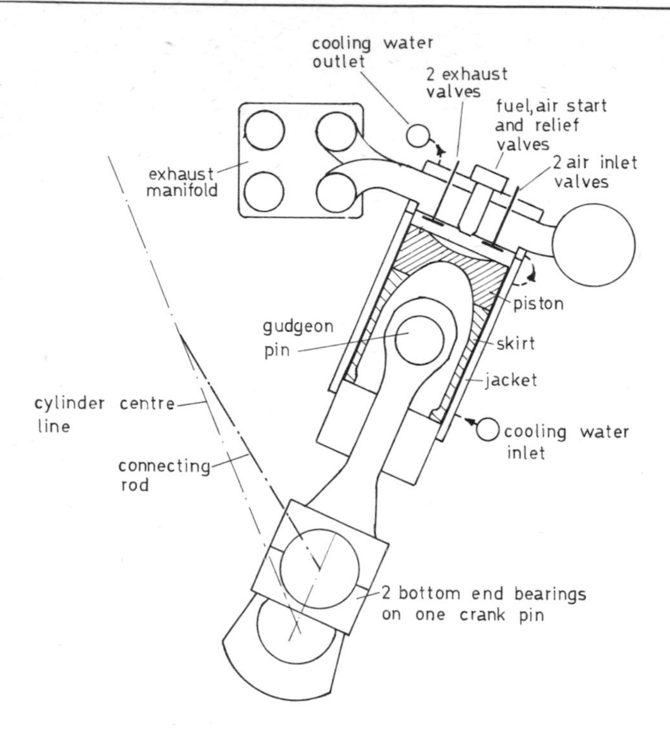

Fig. 106

Future Development

The trend in the field of the medium speed diesel engine is towards higher power outputs per cylinder. About seven engine designs have prototypes under construction or operating on test beds, etc., at the time of writing this book. Some have been ordered for LPG and products tankers, fast refrigerated vessels and container ships. The considerable increase in power has resulted in an increase in size of component parts, hence one advantage, namely, ease of handling has been lost.

The following approximate figures give some indication of the trend.

Cylinder bore	560 mm
Stroke	600 mm
Speed	400 rev/min
Power per cylinder	1000 kW

Overall dimensions of a 12 cylinder 'V':

Length	9.5 m
Height	4.0 m
Fuel consumption	200 g/kw h
Thermal Efficiency	42%

A joint all-British venture is the Doxford Seahorse medium speed diesel that is currently undergoing extensive testing, it is a 1865 kW/cylinder at 300 rev/min opposed piston 2-stroke cycle engine 580 mm bore 1300 mm stroke of all welded fabrication. High expectations are being pinned on this engine and it is hoped that they will eventually be realised and once again we may see Doxford diesels in considerable demand.

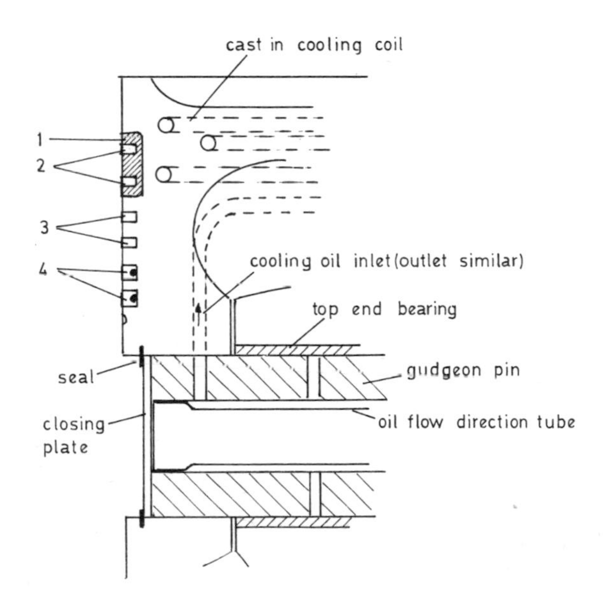

PISTON COOLING

Fig. 107

Typical Lubrication and Piston Cooling System

A pump, which could be main engine driven, supplies oil to a main feeder pipe wherein oil pressure is maintained at approximately 6 bar. Individual pipes supply oil to the main bearings from the feeder, the oil then passes through the drilled crankshaft to the crank pin bearing then flows up the drilled connecting rod to lubricate the small end bush. It then flows around the cooling tubes cast in the piston crown then back down the connecting rod to the engine sump. Oil would also be taken from the main feeder to lubricate camshaft gear drive, camshaft bearings, pump bearings, etc.

Fig. 107 shows in simplified form a typical cooling system for alloy pistons, cast in the piston is a cooling coil and a cast iron ring carrier (marked 1 in the diagram). (2) are two chromium plated compression rings, (3) two copper plated compression rings, (4) two spring backed downward scraping, scraper rings of low inertia type. They are spring backed to give effective outward radial pressure since the gas pressure behind the ring would be very small. The oil flow direction tube is expanded at each end into the gudgeon pin and it is so passaged to direct oil flow and return to their respective places without mixing.

Due to complex vibration problems that can arise in medium speed engines of the 'V' type it would appear important to have a very strong and compact arrangement of bedplate, etc. Excessive vibration of the structure can lead to increased cylinder liner wear and considerable amounts of lubricating oil being consumed. Alkaline lubricating oil of the type used in these engines is expensive and because the engines are mainly trunk type consumption rates can be high. Positioning, and type, of oil scraper ring is important. With some engines they have been moved from a position below the gudgeon pin to above since considerable end leakage sometimes occurred from the gudgeon bearing. The rings should scrape downwards and there may be two scraper rings fitted each with two downward scraping edges, spring backed and of low inertia.

Aluminium alloy pistons could give rise to what is termed incendive sparking. Aluminium in powder form is used in 'Thermit Welding' and it could also be used in flares or incendiaries, this would indicate that it greedily absorbs oxygen, *i.e.* burns easily. Hence if pistons made of aluminium are rubbed, in the dry state, across steel, sparking may occur which in an explosive atmosphere could be very dangerous.

TEST EXAMPLES

First Class

1. Sketch and describe a suitable reduction gear capable of being reversed for use with medium speed diesel engines engaged in main propulsion.

2. Explain the reasons why flexible couplings are fitted to medium speed diesel engines. Sketch and describe two types of coupling.

3. In some medium and high speed diesel engines the cranks are arranged in 'V' configuration instead of an in-line configuration. Why is this? Show with the aid of sketches how the drive is transmitted to the crankshaft.

4. Discuss the advantages and disadvantages of friction and fluid types of clutches for the coupling of medium speed diesel engines to a single propeller.

5. Why are multi-engined medium speed diesel installations preferred to large slow running diesel plants? Sketch a suitable system for main propulsion in which two medium speed diesels are coupled to a single controllable pitch propeller.

6. Discuss the problems associated with exhaust valves for medium speed diesel engines burning heavy fuel oil.

Second Class

1. Sketch and describe a system for main propulsion in which two medium speed diesel engines are coupled to a single propeller.

2. Discuss the advantages and disadvantages of medium speed diesel engines compared to large slow running engines.

3. Sketch and describe a clutch suitable for connecting a medium speed diesel to a power take-off shaft.

4. Sketch and describe an exhaust valve suitable for a medium speed diesel engine.

5. Describe the lubrication and cooling system for the bearings and pistons of a medium speed trunk type of diesel engine.

6. Explain why lubricating oil consumption is greater in medium speed diesels than in slow running diesels and the steps taken to minimise the consumption.

CHAPTER 9

WASTE HEAT PLANT

GENERAL DETAILS

Reference should be made to Chapter 1 for general comments relating to a heat balance. Fig. 3 details an approximate heat balance for an I.C. engine with losses of 35% and 22% to exhaust and cooling respectively. Fig. 4 analyses such losses in more detail. Every attempt must be made to utilise energy in waste heat and recovery from both exhaust gas and coolant is established practice. Sufficient heat energy potential is normally available in exhaust gas at full engine power to supply total electrical and heating services for the ship. The amount of heat actually recovered from the exhaust gases depends upon various factors such as steam pressure, temperature and evaporative rate required; exhaust gas inlet temperature, mass flow of gas, condition of heating surfaces, etc. Waste heat boilers can recover up to about 60% of the loss to atmosphere in exhaust gases. Jacket water temperature is usually restricted to about 70°C and heat recovery is low grade. A generous allowance must be made for parasitic loss, including radiation, in the plant design.

Combustion Equipment

Obviously most boilers and heaters have arrangements for burning oil fuel during low engine power conditions. It is therefore appropriate to repeat some very general remarks on combustion with details of typical equipment in use.

Good combustion is essential for the efficient running of the boiler as it gives the best possible heat release and the minimum amount of deposits upon the heating surfaces. To ascertain if the combustion is good we measure the % CO_2 content (and in some installations the % O_2 content) and observe the appearance of the gases.

If the % CO_2 content is high (or the % O_2 content low) and the gases are in a non smoky condition then the combustion of the fuel is correct. With a high % CO_2 content the % excess air required for combustion will be low and this results in improved boiler efficiency since less heat is taken from the burning fuel by the small amount of excess air. If the excess air supply is increased then the % CO_2 content of the gases will fall.

Condition of burners, oil condition pressure and temperature, condition of air registers, air supply pressure and temperature are all factors which can influence combustion.

Burners: If these are dirty or the sprayer plates are damaged then atomisation of the fuel will be affected.

Oil: If the oil is dirty it can foul up the burners. (Filters are provided in the oil supply lines to remove most of the dirt particles but filters can get damaged. Ideally the mesh in the last filter should be smaller than the holes in the burner sprayer plate.)

Water in the oil can affect combustion, it could lead to the burners being extinguished and a dangerous situation arising. It could also produce panting which can result in structural defects.

If the oil temperature is too low the oil does not readily atomise since its viscosity will be high, this could cause flame impingement, overheating, tube and refractory failure. If the oil temperature is too high the burner tip becomes too hot and excessive carbon deposits can then be formed on the tip causing spray defects, these could again lead to flame impingement on adjacent refractory and damage could also occur to the air swirlers.

Oil pressure is also important since it affects atomisation and lengths of spray jets.

Air Register: Good mixing of the fuel particles with the air is essential, hence the condition of the air registers and their swirling devices are important, if they are damaged mechanically or by corrosion then the air flow will be affected.

Air: The excess air supply is governed mainly by the air pressure and if this is incorrect combustion will be incorrect.

PACKAGE BOILERS

Although such boilers are not necessarily involved with waste heat systems it is considered appropriate to include them at this stage. These boilers are often fitted on motorships for auxiliary use and the principles and practice are a good lead into general boiler practice. Two types of design involving modern principles will now be considered.

Multi-water-tube Vertical Boiler

The design sketched in Fig. 108 is based on the Spencer-Hopwood boiler (English Electric).

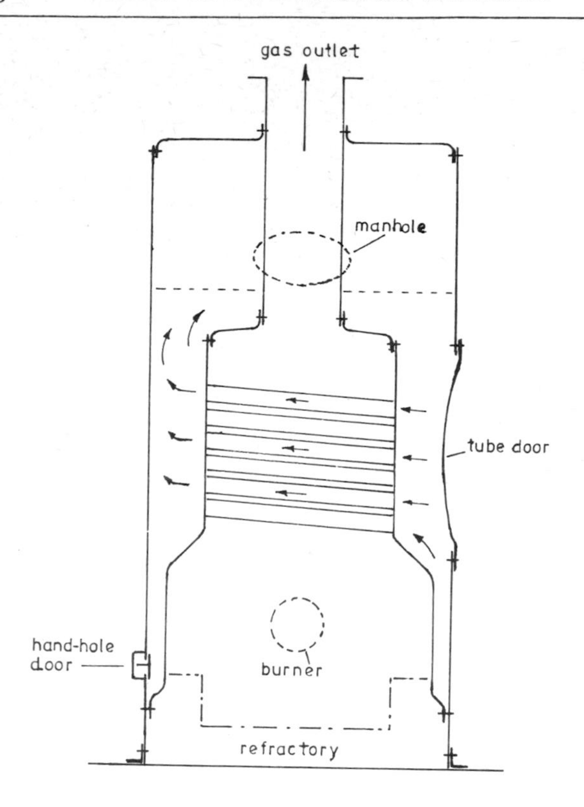

MULTI WATER-TUBE VERTICAL BOILER

Fig. 108

The water space and flue gas path should be clear from Fig. 108. The inclined water tubes promote vigorous circulation with good heat transfer and reduced scale deposits. Water deflector plates are often fitted at tube outlets on automatic boilers to reduce agitation of water surface and to improve circulation. Uptakes are fitted with cast iron liners, doubling plates compensate for manhole and tube door and the firebox vertical half seam is welded. As a package unit a burner, feed pump, combustion and water level control, etc., are incorporated. Chemical dosage and continuous blowdown can be arranged. Some typical data figures for a given size would be: from 1.66 litre/h of fuel, 19 kg/s of steam at 10.5 bar is produced from feed at 16°C with heating surface 27m^2.

Mass of boiler 8760 kg, capacity 1580 litre, height 3785 mm, diameter 1550 mm. Standard mountings are fitted to welded pads and the burner arrangement gives a slightly downward direction of flame pattern.

Vapour Vertical Boiler (Coiled-Tube)

Fig. 109 shows in a simplified diagrammatic form a coiled-tube boiler of the Stone-Vapor type. It is compact, space saving, designed for u.m.s. operation, and is supplied ready for connecting to the ships services.

A power supply, depicted here by a motor, is required for the feed pump, fuel pump (if fitted), fan and controls.

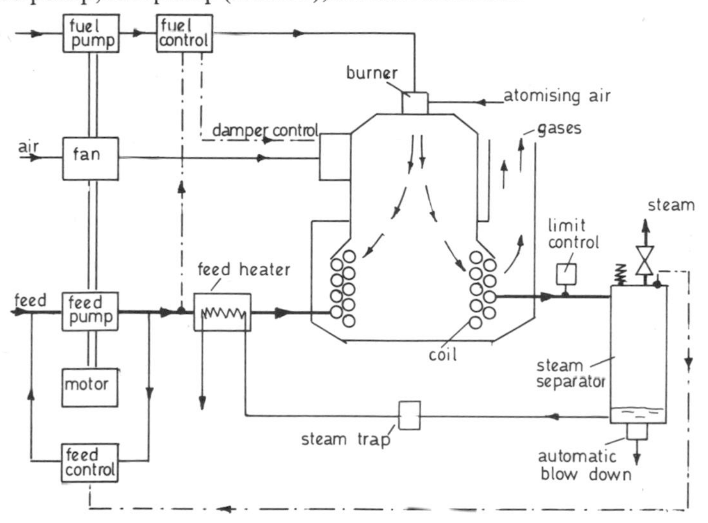

PACKAGE COIL TYPE BOILER
Fig. 109

Feed water is force circulated through the generation coil wherein about 90% is evaporated. The un-evaporated water travelling at high velocity carries sludge and scale into the separator, which can be blown out at intervals manually or automatically. Steam at about 99% dry is taken from the separator for shipboard use.

The boiler is completely automatic in operation. If, for example, the steam demand is increased, the pressure drop in the separator is sensed and a signal, transmitted to the feed controller, demands increased feed, which in turn increases air and fuel supply.

With such a small water content explosion due to coil failure is virtually impossible and a steam temperature limit control protects the coil against abnormally high temperatures. In addition the servo-fuel control protects the boiler in the event of failure of water supply. Performance of a typical unit could be:

Steam pressure	10 bar
Evaporation	3000 kg/h
Thermal efficiency	80%

Full steam output in about 3 to 4 mins.

Note: Atomising air for the fuel may be required at a pressure of about 5 bar.

Steam to Steam Generation

In vessels which are fitted with water tube boilers a protection system of steam to steam generation may be used instead of desuperheaters and reducing valves, etc. (See later.)

TURBO GENERATORS

Such turbines are fairly standard l.p. steam practice and reference, where necessary, could be made to Volume 9. Detailed instructions are provided on board ship for personnel unfamiliar with turbine practice. For the purposes of this chapter the short extract description given below should be typical and adequate.

Turbine

A single cylinder, single axial flow, multistage (say 5) impulse turbine provided with steam through nozzles at 10 bar and 300°C preferably with superheat to limit exhaust moisture to 12%. Axial adjustment of rotor position is usually arranged at the thrust block and protection for overspeed, low oil pressure and low vacuum are provided. Materials and construction for the turbine unit and single reduction gearing are standard modern practice.

Electrical

The turbine at 100-166 rev/s drives the alternator and exciter through a reduction of about 6:1 to produce typically 450-600 kW at 440 V, 3 ph., 60 Hz. A centrifugal shaft-driven motorised governor arranged for local or switchboard operation would operate the throttle valve via a hydraulic servo. Straight line electrical characteristics normally incorporates a speed droop adjustment to allow ready load sharing with auxiliary diesel generators or an extra turbo unit.

Ancillary Plant

This is normally provided as a package unit with condenser, air ejector, auto gland seals, gland condenser, motorised and worm-driven oil pumps, etc. A feed system is provided either integral or divorced from the turbine-gearbox-alternator unit. Exhaust can be arranged to a combined condenser incorporating cargo exhaust. Control utilises gas by-pass, dumping steam, etc.

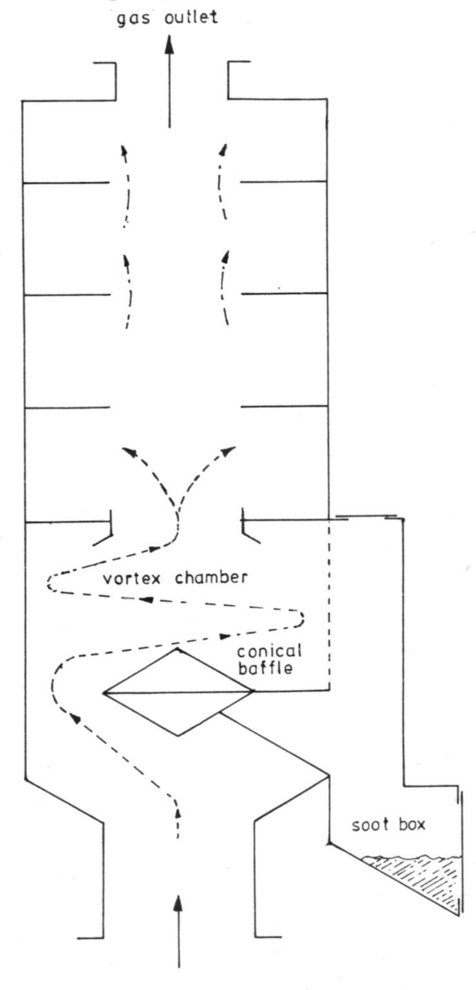

Fig. 110

SILENCERS

Normally waste heat boilers act as spark arresters and silencers at all times. The silencer sketched in Fig. 110 would not usually be fitted if such boilers were used but a short description of the silencer may be useful.

Three designs have been utilised. The tank type has a reservoir of volume about 30 times cylinder volume. Baffles are arranged to give about four gas reversals. The diffuser type has a central perforated discharge pipe surrounded by a number of chambers of varying volume. The orifice type is sketched in Fig. 110 and the construction should be clear. Energy pulsations and sound waves are dissipated by repeated throttling and expansion.

GAS ANALYSIS

A number of factors have been stated which affect the design and operation of the plant and some salient points will now be briefly considered.

Optimum Pressure

This depends on the system adopted but in general the range is from 6 bar to 11 bar. The lower pressures give a cheaper unit with near maximum heat recovery. However higher pressures allow more flexibility in supply with perhaps more useful steam for certain auxiliary functions together with reserve steam capacity to meet variations in demand. Low feed inlet temperatures reduce pressure and evaporative rate.

Temperature

A minimum temperature differential obviously applies for heat transfer. Temperature difference, fouling, gas velocity, gas distribution, metal surface resistance, etc., are all important factors. Reduction in service engine revolutions will cause reduced gas mass and temperature increase if the power is maintained constant. A similar effect will be apparent under operation in tropical conditions. The effect of increased back pressure will be to raise the gas temperature for a given air inlet temperature. Fig. 111 illustrates: (a) typical heat transfer diagram, and (b) gas temperature/mass-power curves. A common temperature differential is about 40°C, *i.e.* water inlet 120°C and gas exit 160°C.

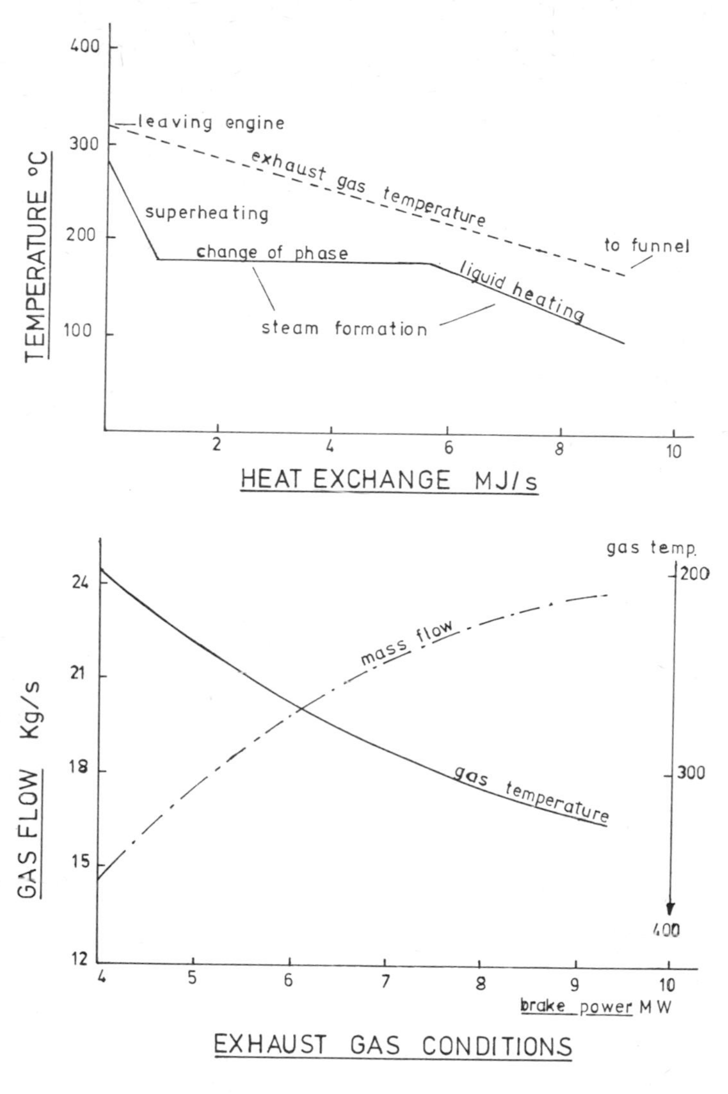

Fig. 111

Corrosion

The acid dew point expected is about 110°C with a 3% sulphur fuel and a high rate of conversion from SO_2 to SO_3 is possible. Minimum metal temperatures of 120°C for mild steel are required.

Exhaust System

The arrangement must offer unrestricted flow for gases so that back pressure is not increased. Good access is required for inspection and cleaning. On designs with alternate gas-oil firing provision must be made for quick and foolproof change-over with no possibility of closure to atmosphere and waste heat system at the same time.

GAS/WATER HEAT EXCHANGERS

Waste Heat Economisers

Such units are well proven in steamship practice and similar all-welded units are reliable and have low maintenance costs in motorships. Gas path can be staggered or straight through with extended surface element construction. Large flat casings usually require good stiffening against vibration. Water wash and soot blowing fittings may be provided.

Waste Heat Boilers

These boilers have a simple construction and fairly low cost. At this stage a single natural circulation boiler will be considered and these normally classify into three types, namely: simple, alternate and composite.

Simple

These boilers are not very common as they operate on waste heat only. Single or two-pass types are available, the latter being the most efficient. Small units of this type have been fitted to auxiliary oil engine exhaust systems, operating mainly as economisers, in conjunction with another boiler. A gas change valve to direct flow to the boiler or atmosphere is usually fitted as described below.

Alternate

This type is a compromise between the other two. It is arranged to give alternate gas and oil firing with either single or double pass gas flow. It is particularly important to arrange the piping system so that oil fuel firing is prevented when exhaust gas is passing through the boiler. A large butterfly type of change-over valve is fitted before the boiler so as to direct exhaust gas to the boiler or to the atmosphere. The valve is so arranged that gas flow will not be obstructed in that as the valve is closing one outlet the other outlet is being opened. The operating mechanism, usually a large external square thread, should be arranged so that with the valve directed to the boiler, fuel oil is shut off. A mechanical system using an extension piece can be arranged to push a fork lever into the operating handwheel of the oil fuel supply valve. When the exhaust valve is fully operated to direct the gas to atmosphere the fork lever then clears the oil fuel valve handwheel after change-over travel is completed. It is also very important to ensure full fan venting and proper fuel heating-circulation procedures before lighting the oil fuel burners.

Composite

Such boilers are arranged for simultaneous operation on waste heat and oil fuel. The oil fuel section is usually only single pass. Early designs utilised Scotch boilers, with, say, a three furnace boiler, it may mean retaining the centre or the wing furnaces for oil fuel firing. The gas unit would often have a lower tube bank in place of the furnace, with access to the chamber from the boiler back, so giving double pass. Alternative single pass could be arranged with gas entry at the boiler back. Exhaust and oil fuel sections would have separate uptakes and an inlet change-over valve was required. In general Scotch boilers as described are nearly obsolete and vertical boilers are used. As good representative, and more up-to-date, common practice, two types of such boiler will be considered.

Cochran Boiler

The Cochran boiler whose working pressure is normally of the order of 8 bar is available in various types and arrangements, some of which are:

Single pass composite, *i.e.* one pass for the exhaust gases and two uptakes, one for the oil fired system and one for exhaust system. Double pass composite, *i.e.* two passes for the exhaust gases and two uptakes, one for the oil fired system and one for the exhaust system. (Double pass exhaust gas, no oil fired furnace and a single uptake, is available as a simple type. Or, double pass alternatively fired, *i.e.* two passes from the furnace for either exhaust gases or oil fired system with one common uptake).

The boiler is made from good quality low carbon open hearth mild steel plate. The furnace is pressed out of a single plate and is therefore seamless.

Connecting the bottom of the furnace to the boiler shell plating is a seamless 'Ogee' ring. This ring is pressed out of thicker plating than the furnace, the greater thickness is necessary since circula-

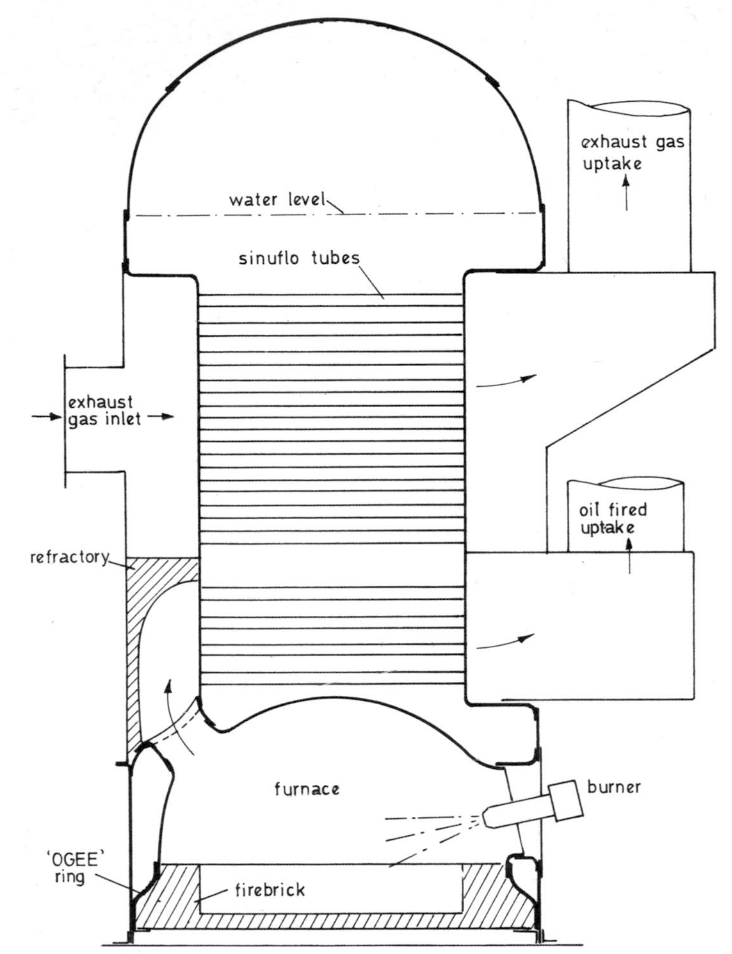

DIAGRAMMATIC ARRANGEMENT OF A SINGLE PASS COMPOSITE COCHRAN BOILER

Fig. 112

tion in its vicinity is not as good as elsewhere in the boiler and deposits can accumulate between it and the boiler shell plating. Hand hole cleaning doors are provided around the circumference of the boiler in the region of the 'Ogee' ring.

The tube plates are supported by means of the tubes and by gusset stays, the gusset stays supporting the flat top of the tube plating.

Tubes fitted, are usually of special design (Sinuflo), being smoothly sinuous in order to increase heat transfer by baffling the gases. The wave formation of the tubes lies in a horizontal plane when the tubes are fitted, this ensures that no troughs are available for the collection of dirt or moisture. This wave formation does not in any way affect cleaning or fitting of the tubes.

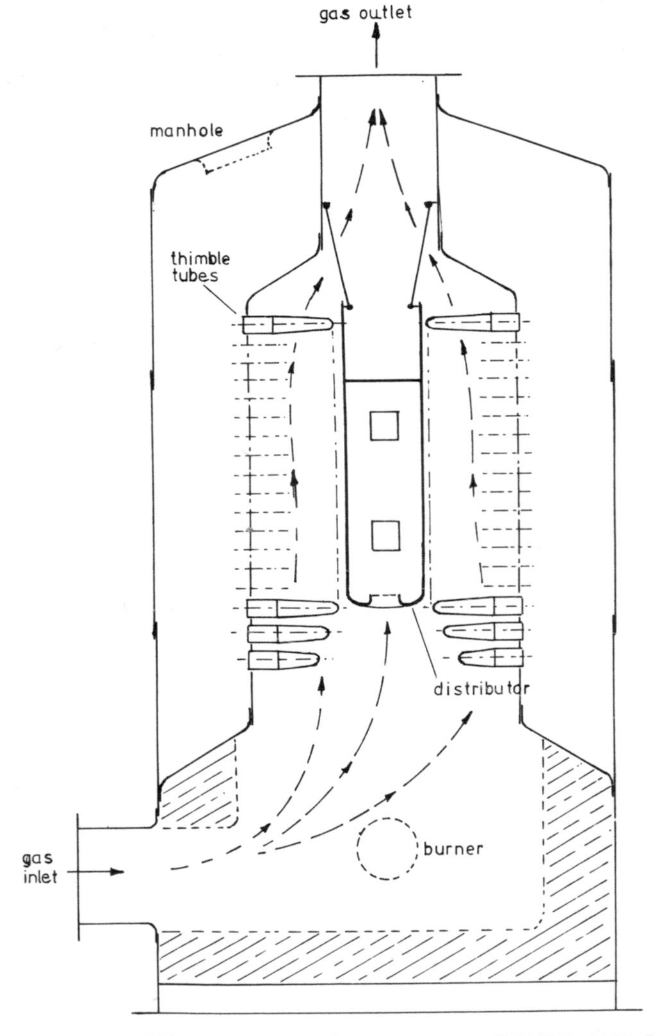

ALTERNATIVELY FIRED THIMBLE-TUBE BOILER

Fig. 113

Thimble Tube Boiler

There are various designs of thimble tube boiler, these include: oil fired, exhaust gas, alternatively fired and composite types.

The basic principle with which the thimble tube operates was discovered by Thomas Clarkson. He found that a horizontally arranged tapered thimble tube, when heated externally, could cause rapid ebullitions of a spasmodic nature to occur to water within the tube, with subsequent steam generation.

Fig. 113 shows diagrammatically an alternatively fired boiler of the Clarkson thimble tube type capable of generating steam with a working pressure of 8 bar. The cylindrical outer shell encloses a cylindrical combustion chamber, from which, radially arranged thimble tubes project inwards. The combustion chamber is attached to the bottom of the shell by an 'Ogee' ring and to the top of the shell by a cylindrical uptake. Centrally arranged in the combustion chamber is an adjustable gas baffle tube.

EXHAUST GAS HEAT RECOVERY CIRCUITS

Many circuits are possible and a few arrangements will now be considered. Single boiler units as discussed, whilst cheap, are not flexible and have relatively small steam generating capacity. The systems now considered are based on multi-boiler installations.

Natural Circulation Multi-Boiler System

It is possible to have a single exhaust gas boiler located high up in the funnel, operating on natural circulation whereby a limited amount of steam is available for power supply whilst the vessel is at sea. In port or during excessive load conditions, the main boiler or boilers are brought into operation to supply steam to the same steam range by suitable cross connecting steam stop valves. In port the exhaust gas boiler is secured and all steam supplied by the oil fired main boilers. This system is suitable for use on vessels such as tankers where a comparatively large port steaming capacity may be required for operation of cargo pumps, but suffers from the disadvantage that the main boilers must either be warmed through at regular intervals or must be warmed through prior to reaching port. Further to this the main boilers are not immediately ready for use in event of an emergency stop at sea unless the continuous warming through procedure has been followed

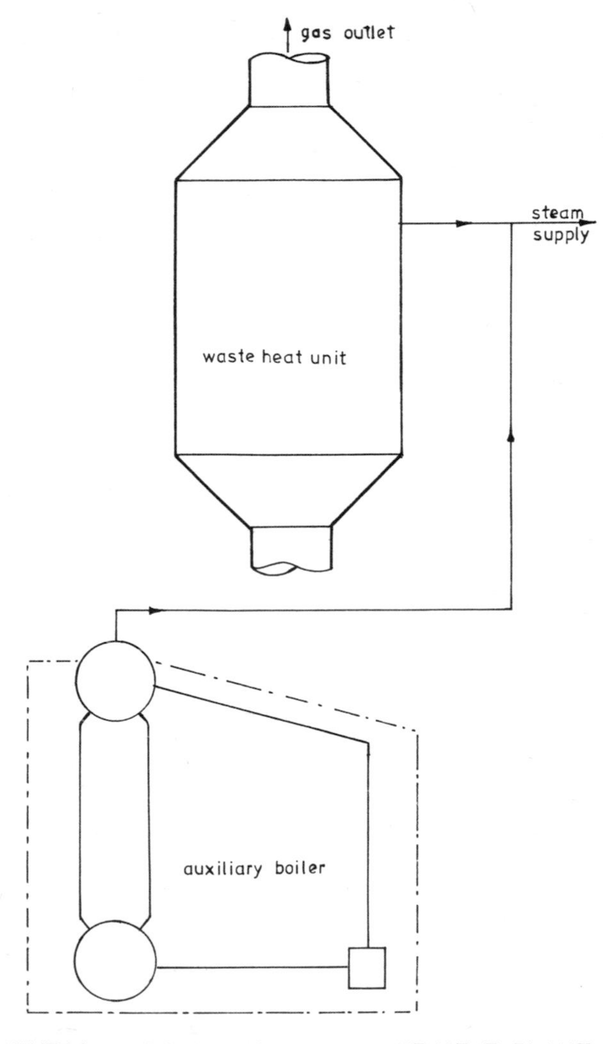
gas outlet
steam supply
waste heat unit
auxiliary boiler
NATURAL CIRCULATION
WASTE HEAT PLANT AND W.T. BOILER

Fig. 114

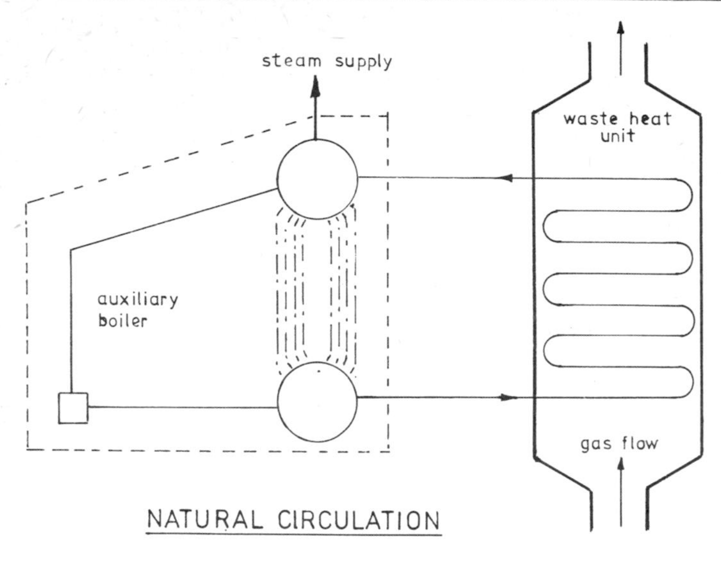

Fig. 115

Forced Circulation Multi-boiler System

In order to overcome the difficulties mentioned above, and to render more efficient heat transfer possible, a simple forced circulation system may be employed. The exhaust gas boiler is arranged to be a drowned heat exchanger which, due to the action of a circulating pump, discharges its steam and water emulsion to the steam drum of a water-tube boiler. The forced circulation pump draws from near the bottom of the main boiler water drum and circulates water at almost ten times the steam production rate so giving good heat transfer. The steam/water emulsion on being discharged into the water space of the main boiler drum separates out exactly in the same way as if the boiler were being oil fired. This arrangement ensures that the main boiler is always warm and capable of being immediately fired by manual operation or supplementary pilot operated automatic fuel burning equipment. Feed passes to the main boiler and becomes neutralised by chemical water treatment. Surface scaling is thus largely precluded and settled out impurities can be removed at the main boiler blowdown. If feed flow only is passed through an economiser type unit parallel flow reduces risks of vapour locking. Unsteady feed flow at normal gas conditions can result in water flash over to steam and rapid metal temperature variations. Steam, hot water and cold water conditions can cause thermal shock and water hammer. Contra flow designs are generally more efficient from a heat transfer viewpoint giving gas temperatures nearer steam temperatures and are certainly preferred for economisers if circulation rate is a multiple of feed flow. The generation section is normally parallel flow and the superheat section contra flow. Output control could be arranged by output valves at two different levels so varying the effective heat transfer surface utilised. In addition a circulating pump by-pass arrangement gives an effective control method.

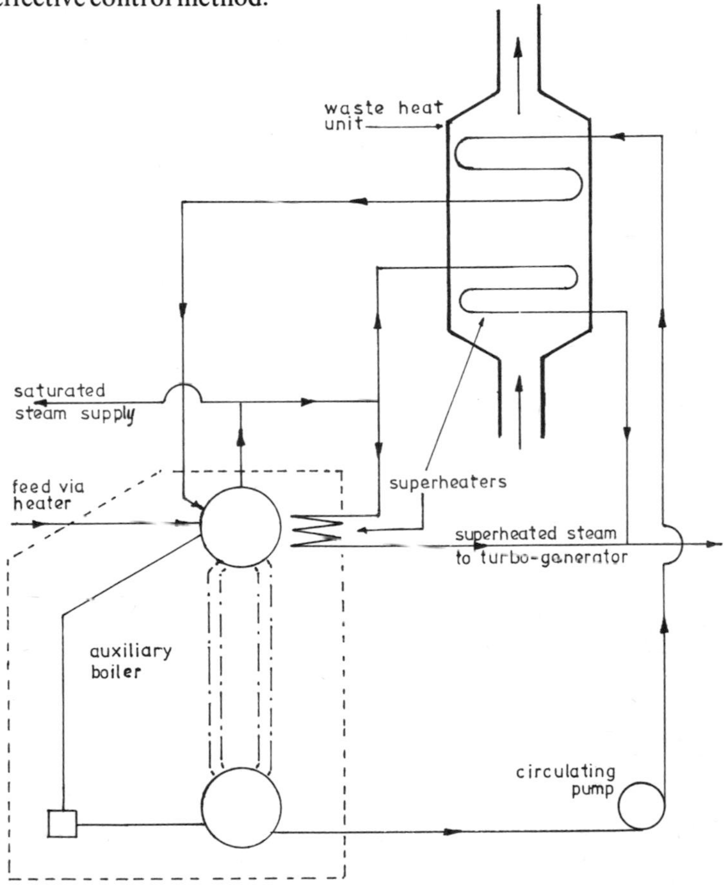

Fig. 116

Dual Pressure Forced Circulation Multi-Boiler System

This concept has been incorporated in the latest waste heat circuits and the sketch illustrates how the general principle can be applied in conjunction with a waste heat exchanger to supply superheated steam. By this means every precaution has been taken to minimise the effect of contamination of the water-tube boiler.

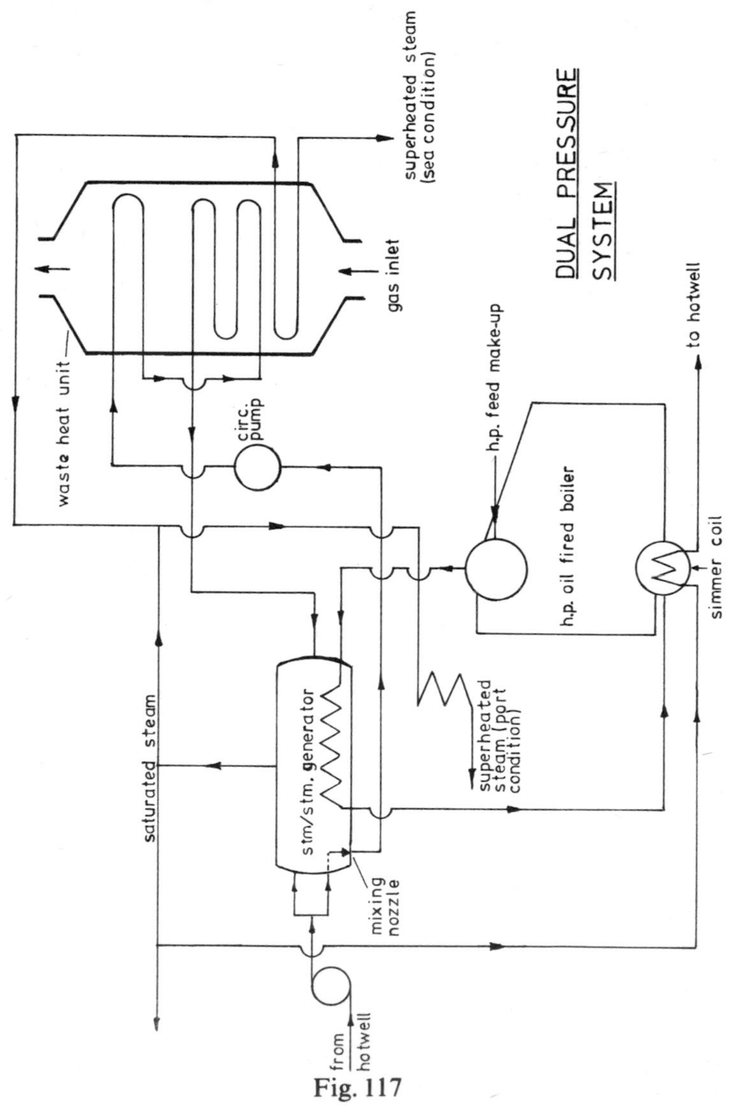

Fig. 117

Steam generated in the water-tube boiler by either oil firing or waste heat exchanger passes through a submerged tube nest in the steam/steam generator to give lower grade steam which is subsequently passed to the superheater.

A water-tube boiler, steam/steam generator and feed heater may be designed as a packaged unit with the feed heater incorporated in the steam/steam generator. The high pressure high temperature system at say 10 bar will supply a turbo generator for all electrical services whilst the low pressure system at say 2½ bar would provide all heating services. Obviously the dual system is more costly. Numerous designs are possible including separate lp and hp boilers, either natural or forced circulation, indirect systems with single or double feed heating, etc.

WATER/WATER HEAT EXCHANGERS

Evaporators

The basic information given on evaporators in Volume 8 should *first* be considered. In motor ship practice efficient single effect units incorporating flexible elements and controlled water level are in service. Outputs of 0.5 tonne per day (exhaust steam) and 1.0 tonne per day (live steam) are common. Flash evaporators have increasingly been fitted on large vessels utilising multi-stage units. Also multiple effect evaporators of conventional form are used. The steam circuit of many modern motor ships has developed in complexity to approach successful steam ship practice.

FEED HEATING

The advantages of pre-heating feed water are obvious. Three methods will be considered, namely: economisers, mixture and indirect. Economiser types have been included in previous discussion and sketches. It is sufficient to repeat that such systems require a careful design to cope with fluctuations of steam demand and that particular attention is necessary to ensure protection against corrosive attack. Mixture systems employ parallel feeding with circulating pump and feed pump to the economiser inlet. Such circuits require careful matching of the two pumps and control has to be very effective to prevent cold water surges leading to reducing metal temperatures and causing corrosion. Indirect systems require a water/water exchanger feed heater.

This design reduces the risk of solid deposit in the economiser and maintains steady conditions of economiser water flow so protecting the economiser against corrosive attack. A typical system is shown in Fig. 118. If boiler pressure tends to rise too high the

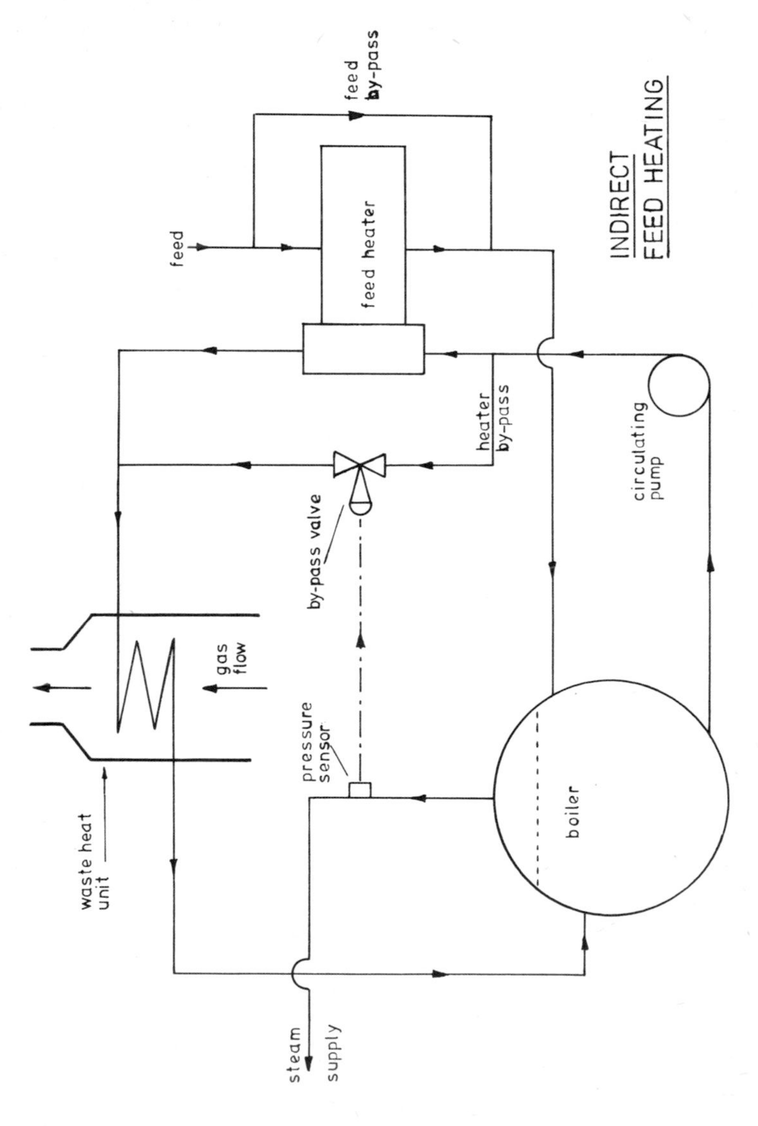

Fig. 118

circulation by-pass will be opened. The effect will be twofold, *i.e.* feed water will enter the boiler at the lower temperature and water temperature entering the economiser is at a higher temperature. These two effects serve to reduce boiler pressure and so control the system. Obviously this system is more costly but is very flexible.

COMBINED HEAT RECOVERY CIRCUITS

The low grade heat of engine coolant systems restricts the heat recovery in such secondary circuits to temperatures near 60°C. As such it is normally restricted to use with distillation plants. Combined or compound units involving combination between engine coolant and exhaust gas systems are complicated by the need to prevent contamination and utilise the large volume of low temperature coolant in circulation. Jacket water coolant temperatures have increased in recent motorship practice but even if the engine design can be modified to suit even higher temperatures there is always a problem of high radiation heat loss from jackets to confined engine rooms.

TEST EXAMPLES

Second Class

1. Sketch diagrammatically and describe a boiler which may be alternatively heated by oil fuel burning or exhaust gas in which the heat transfer surfaces are common. Describe in detail the change-over arrangements and mention any safety devices fitted to this gear.

2. Sketch and describe a thimble tube type of boiler arranged for heat reception from oil fuel burning or exhaust gas. Discuss working pressure and method of operation.

3. With the aid of a sketch describe an auxiliary boiler which uses exhaust gas and oil firing simultaneously. Enumerate the fittings and mountings required.

4. Sketch and describe a boiler which utilises exhaust gas. Give the gas temperature entering and leaving and state the boiler pressure.

5. Discuss explosions which occur in uptakes and silencers. These often occur upon re-starting the engine after it has been stopped for a short period. What are the causes and how can they be minimised?

First Class

1. Sketch and describe a composite thimble tube boiler. Describe how the thimble tubes are fitted and discuss burning of tube ends and other possible defects.

2. Sketch and describe an exhaust gas boiler worked in conjunction with a large oil engine. On an inspection of such a boiler what particular defects would you look for? How often should such a boiler be opened up for inspection?

3. Sketch and describe a forced circulation boiler suitable for producing saturated steam at 8 bar.

4. Sketch and describe in detail a furnace front for an auxiliary boiler and show all the fittings. Describe in sequence how to raise steam on this boiler starting from cold.

5. Sketch and describe a waste heat system suitable for use on a large modern tanker. Discuss the units and briefly describe their functions, in relation to the system design.

CHAPTER 10
MISCELLANEOUS

CRANKCASE EXPLOSIONS

Introduction

The student should first refer to Volume 8 for a consideration of spontaneous ignition temperatures and also limits of inflammability in air. Crankcase explosions have occurred steadily over the years with perhaps that of the *Reino-del-Pacifico* in 1947 the most serious of all. In fact crankcase explosions have occurred in all types of enclosed crankcase engines, including steam engines. Explosions occur in both trunk piston types and in types with a scraper gland seal on the piston rod. Much research has been done in this field but the difficulties of full experimentation utilising actual engines under normal operating conditions is almost impossible to attempt. The following is a simplified presentation based on the mechanics of cause of explosion, appropriate DoT regulations and recommendations and descriptive details of preventative and protective devices utilised.

Mechanics of Explosion

1. A hot spot is an essential source of such explosions in crankcases as it provides the necessary ignition temperature, heat for oil vapourisation and possibly ignition spark. Normal crankcase oil spray particles are in general too large to be easily explosive (average 200 microns). Vapourised lubricating oil from the hot source occurs at 400°C, in some cases lower, with a particle size explosive with the correct air ratio (average 6 microns). Vapour can condense on colder regions, a condensed mist with fine particle size readily causes explosion in the presence of an ignition source. A lower limit of flammability of about 50 mg/l is often found in practice. Experiment indicates two separate temperature regions in which ignition can take place, *i.e.* 270°C-350°C and above 400°C.

2. Initial flame speed after mist ignition is about 0.3 m/s but unless the associated pressure is relieved this will increase to about 300 m/s with corresponding pressure rise. In a long crankcase, flame speeds of 3 km/s are possible giving detonation and maximum damage. The pressure rise varies with conditions but

without detonation does not normally exceed 7 bar and may often be in the range 1-3 bar.

3. A primary explosion occurs and the resulting damage may allow air into a partial vacuum. A secondary explosion can now take place, which is often more violent than the first, followed by similar sequence until equilibrium.

4. The pressure generated, as considered over a short but finite time, is not too great but instantaneously is very high. The associated flame is also dangerous. The gas path cannot ordinarily be deflected quickly due to the high momentum and energy.

5. Devices of protection must allow gradual gas path deflection, give instant relief followed by non return action to prevent air inflow and be arranged to contain flame and direct products away from personnel.

6. Delayed ignition is sometimes possible. An engine when running with a hot spot may heat up through the low temperature ignition region without producing flame because of the length of ignition delay period at lower temperatures. Vapourised mist can therefore be present at 350-400°C. If the engine is stopped the cooling may induce a dangerous state and explosion. Likewise air ingress may dilute a previously too rich mixture into one of dangerous potential.

7. Direct detection of overheating by thermometry offers the greatest protection but the difficulties of complete surveillance of all parts is prohibitive.

8. A properly designed crankcase inspection door preferably bolted in place, suitably dished and curved with say a 3 mm thickness of sheet steel construction should withstand static pressures up to 12 bar although distorted.

9. There are many arguments for and against vapour extraction by exhauster fans. There is no access of free air to the crankcase and the fan tends to produce a slight vacuum in the crankcase. On balance most opinion is that the use of such fans can reduce risk of explosion. The danger of fresh air drawn in to an existing over rich heated state is obvious. On the practical aspect leakage of oil is reduced.

Crankcase Safety Arrangements

The following are based on specific DoT rules:

1. Means should be adapted to prevent danger from the result of explosion.

2. Crankcases and inspection doors should be of robust construction. Attachment of the doors to the crankcase (or entablature) should be substantial.

3. One or more non-return pressure relief valves should be fitted to the crankcase of each cylinder and to any associated gearcase.

4. Such valves should be so arranged or their outlets so guarded that personnel are protected from flame discharged with the explosion.

5. The total clear area through the relief valves should not normally be less than 9.13 cm^2/m^3 of gross crankcase volume.

6. Engines not exceeding 300 mm cylinder bore with strongly constructed crankcases and doors may have relief valve or valves at the ends only. Similarly constructed engines not exceeding 150 mm cylinder bore need not be fitted with relief valves.

7. Lubricating oil drain pipes from engine sump to drain tank should extend to well below the working oil level in the tank.

8. Drain or vent pipes in multiple engine installations are to be so arranged so that the flame of an explosion cannot pass from one engine to another.

9. In large engines having more than six cylinders it is recommended that a diaphragm should be fitted at near mid length to prevent the passage of flame.

10. Consideration should be given to means of detection of overheating and injection of inert gas.

The above should be self explanatory in view of the previous comments made.

Preventative and Protection Devices

In general three aspects are worthy of consideration, *i.e.* relief of explosion, flame protection and explosive mist detection.

Crankcase Explosion Door

A design is shown in Fig. 119. The sketch illustrates a combined valve and flame trap unit with the inspection door insertion in the middle. The internal section supports the steel gauze element and the spider guide and retains the spindle. The external combined aluminium valve and deflector has a synthetic rubber seal. Pressure setting on such doors is often 1/15 bar (above atmospheric pressure). Relief area and allowable pressure rise vary with the licensing insurance authority but a metric ratio of 1:90 should not normally be exceeded based on gross crankcase volume and this should not allow explosion pressures to exceed about 3 bar.

Flame trap

Such devices are advisable to protect personnel. The vented gases can quickly be reduced in temperature by gauze flame traps from say 1500°C to 250°C in 0.5 m. Coating on the gauzes, greases or engine lubricating oil, greatly increases their effectiveness. The best location of the trap is inside the relief valve when it gives a more even distribution of gas flow across its area and liberal wetting with lubricating oil is easier to arrange. A separate oil supply for this action may be necessary. The explosion door in Fig. 119 has an internal mesh flame trap fitted.

Flame traps effectively reduce the explosion pressure and prevent two stage combustion. Gas-vapour release by the operation of an oil wetted flame trap is not usually ignitable. Typical gauze mild steel wire size is 0.3 mm with 40% excess clear area over the valve area.

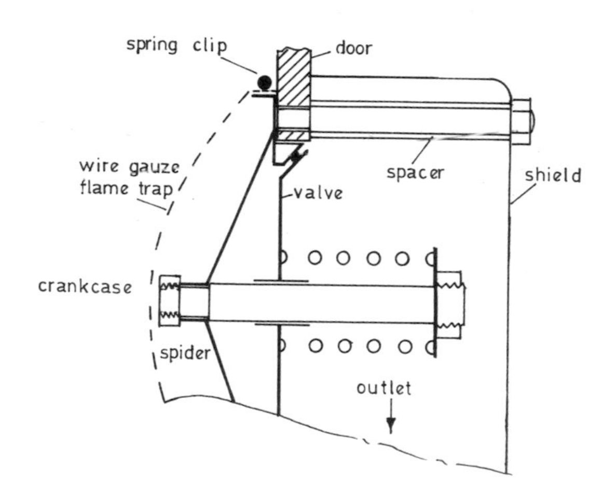

CRANKCASE EXPLOSION RELIEF DOOR

Fig. 119

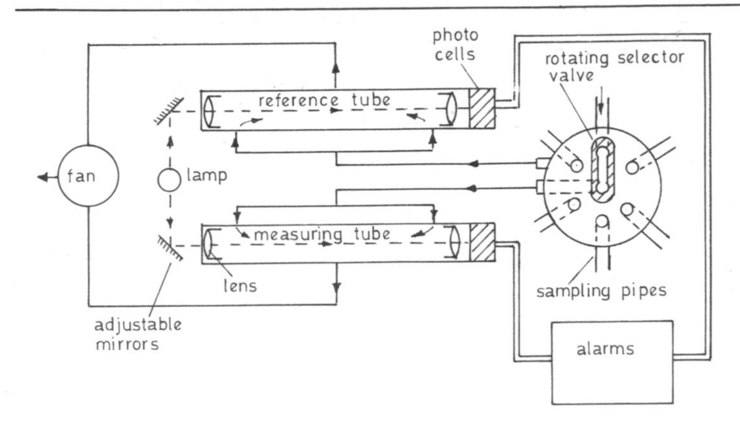

CRANKCASE OIL MIST DETECTOR

Fig. 120

Crankcase Oil Mist Detector

If condensed oil mists are the sole explosive medium then photo-electric detection should give complete protection but if the crankcase spray is explosive the mist detection will only indicate a potential source of ignition. The working of one design of detector should be fairly clear from Fig. 120. The photo cells are normally in a state of electric balance, *i.e.* measure and reference tube mist content in equilibrium. Out of balance current due to rise of crank-case mist density can be arranged to indicate on a galvanometer which can be connected to continuous chart recording and auto visual or audible alarms. The suction fan draws a large volume of slow moving oil-air vapour mixture in turn from various crank-case selection points. Oil mist near the lower critical density region has a very high optical density. Alarm is normally arranged to operate at 2½% of the lower critical point, *i.e.* assuming 50 mg/l as lower explosive limit then warning at 1.25 mg/l.

Operation

The fan draws a sample of oil mist through the rotary valve from each crankcase sampling pipe in turn, then through the measuring tube and delivers it to atmosphere. An average sample is drawn from the rotary valve chamber through the reference tube and delivered to atmosphere at the same time. In the event of overheating in any part of the crankcase there will be a difference in optical density in the two tubes, hence less light will fall on the photo cell in the measuring tube. The photo cell outputs will be different and when the current difference reaches a pre-determined value an alarm signal is operated and the slow turning rotary valve stops, indicating the location of the overheating.

Normal oil particles as spray are precipitated in the sampling tubes and drain back into the crankcase.

CO_2 Drenching System

30% by volume of this inert gas is a complete protection against crankcase explosion. This is particularly beneficial during the dangerous cooling period. Automatic injection can be arranged at, say, 5% of critical lower mist density but in practice many engineers prefer manual operation. When the engine is opened up for inspection, and repair at hot source, it will of course be necessary to ensure proper venting before working personnel enter the crankcase.

GAS TURBINES

The gas turbine theoretical cycle and simple circuit diagram have been considered in Chapter 1. Marine development of gas turbines stemmed from the aero industry in the 1940's. Apart from an early stage of rapid progress the application to marine use has been relatively slow until recently. Consideration can best be applied in three sections, namely, industrial gas turbines, free piston units and aero-derived types. In general this can largely be considered as a 'marinisation' of equipment originally designed for other duty.

Industrial Gas Turbines

The simplest design is a single-shaft unit which has low volume and light weight (5 kg/kW at 20,000 kW). Fuel consumption (specific) may be about 0.36 kg/kWh on residual fuels. This consumption is not normally acceptable for direct propulsion and initial usage was as emergency generators in MN practice and the RN for small vessels or as boost units in larger warships. Compared to steam turbines (32% output, 58% condenser loss) the simple gas turbine (24% output, 73% exhaust loss) is less efficient but the addition of exhaust gas regeneration gives 31% output (specific fuel consumption 0.28 kg/kWh) and combined RN units 36% output. Normally a two-shaft arrangement was preferred in MN practice in which load shaft and compressor shaft are independent.

A design was available by 1955 for main propulsion with maximum turbine inlet conditions of 6 bar, 650°C and specific fuel consumption approaching 0.3 kg/kWh. Starting of the twin-shaft unit was by electric motor, power variation by control of gas flow, conventional gear reduction and propeller drive by hydraulic clutch with astern torque converter (more modern practice uses variable pitch propeller). Turbine and turbo-compressor design utilised standard theory and simple module construction utilising horizontally split casings, diffusers, etc., and easily accessible nozzles.

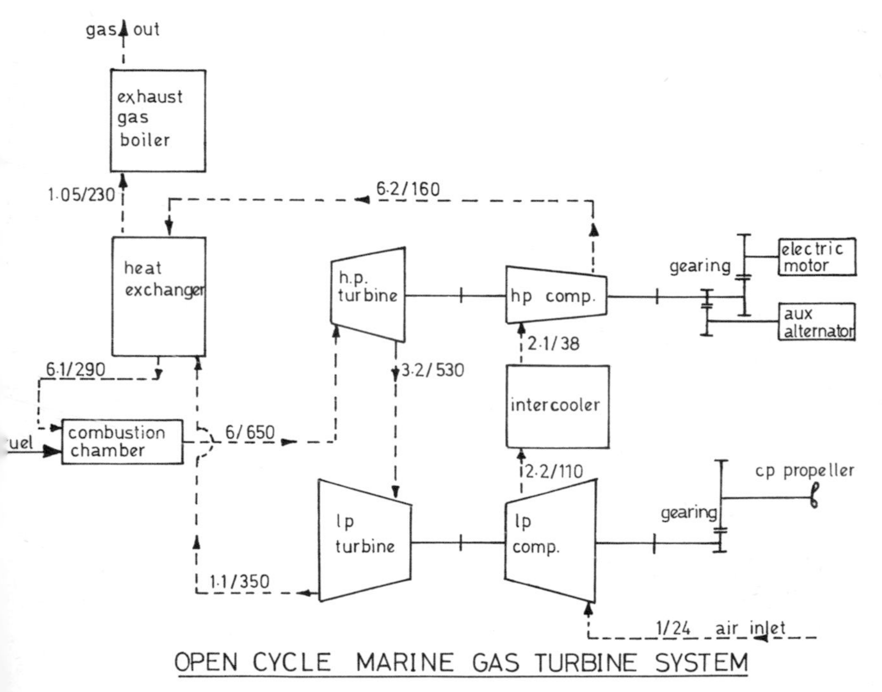

Fig. 121

To improve efficiency even further it is necessary to use much higher inlet gas temperatures (1200°C would give a specific fuel consumption of about 0.2 kg/kWh). The limiting factor is suitable materials. Experiments have been, and still are, being carried out with ceramic blades and with cooled metallic blades. Essentially the problem is the same for steam turbine plant and there has been no marked incentive for the shipowner to install gas turbine plant in preference to equally economic and established steam systems. During the 1960's experience was established in the vessels *Auris, John Sergeant* and *William Paterson.* It may well be that direct gas cooled reactors in conjunction with closed cycle gas turbines in electric power generation may be an attractive possibility in Nuclear technology. G.E.C. produce a wide range (4000-50,000 kW) of industrial gas turbines now effectively marinised for marine propulsion. In addition to reliability, easy maintenance, low volume, etc., the very easy application to electric drives and to automation make the units attractive. Geared drive usually utilises locked train helical gears or alternatively epicyclic gearing.

C.P. propeller development has also broadened the possibilities of various propulsion systems, including geared diesel — gas turbine systems. Marine gas turbines do run with a high noise level and they require to be water washed at regular intervals, the latter depending upon the type of fuel being used. The recently changing design of ships has meant that the owner, or operator, needs to analyse propulsion systems carefully for all economic factors, which vary greatly for VLCC, Ro-Ro, LNG, container vessels, etc. Gas turbines have been exclusively adopted for RN surface vessels.

Free Piston Gas Generator System

This system was installed in a number of vessels built in the decade beginning 1960. The plant consists of gas generators supplying gas to a turbine.

The turbine, geared down and reversible by clutch or variable pitch propeller, is controlled by gas flow. The generators are either shut down or the gas is diverted during manoeuvring. One advantage of the system is flexibility in both machinery layout and the fact that a generator can be taken out of service for overhaul, without seriously reducing engine power. another advantage is that gas temperature is moderate, so reducing material problems, and turbine fouling is reduced. Against this the gas generators do no shaft work, being solely gas producers, and the efficiency of the system is therefore reduced.

Fig. 122 shows a free piston gas generator, it consists of two horizontally opposed piston units reciprocating freely without a crankshaft, in a combined compressor-combustion chamber. At the start of the in-stroke the large air pistons (A) have maximum air pressure in the cushion chamber (E) and move inwards compressing atmospheric pressure air in the compression chamber (F) and receiver (G). Suction valves (H) will be closed

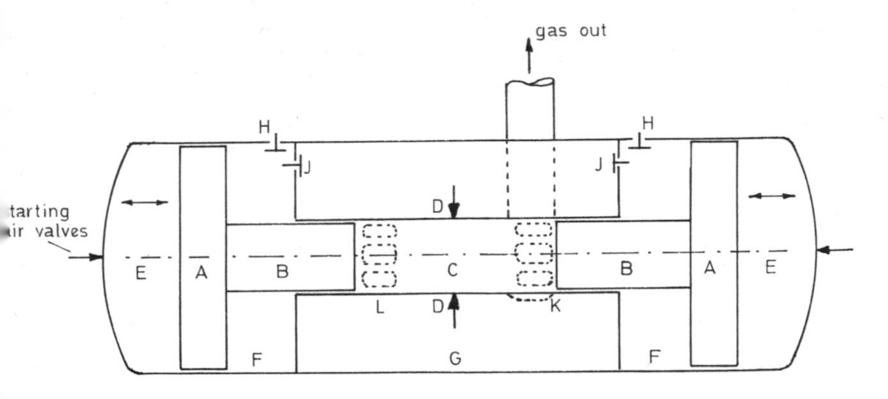

Fig. 122

and discharge valves (J) open. The small pistons (B) compress air in the cylinder (C) to a high pressure and temperature, fuel injectors (D) supply fuel which combusts. Piston units now commence the out stroke.

Air is drawn from the atmosphere into chamber (G) via valves (H), valves (J) closed, air in chambers (E) being compressed. The exhaust ports K will become uncovered and exhaust will commence. Scavenging will occur from chamber (G) into the cylinder via ports (L) as (A) and (B) move to outer extremity.

In the steady state engine cylinder work equals compressor cylinder work. Increasing fuel quantity causes higher pressures and increased effective stroke, with increased frequency of oscillation and gas mass delivery per unit time. Compression ratio is therefore variable with load, and injection timing must be adjusted accordingly. Control links effective stroke of engine to effective stroke of fuel pump. For a moderate generator size about six units give about 7000 kW.

Aero-derived

Apart from RN units so derived from aero gas turbines the first British MN vessel so engined was the g.t.s. *Euroliner* in 1970. Turbo Power and Marine Systems Inc. twin gas turbines, 22,500 kW each at 3600 rev/min drive separate screw shafts at 135 rev/min through double reduction locked train gears. With controllable pitch propellers. Main electrical alternators are driven off the gearbox.

MISCELLANEOUS SPECIMEN QUESTIONS

Second Class

1. State the causes of crankcase explosions. What precautions can be taken at sea and in port to minimise the dangers from such explosions?

2. Sketch and describe a free piston gasifier and list the advantages claimed for the system.

3. Discuss the blowing down and opening up of an auxiliary boiler for cleaning and maintenance. State what points require particular attention.

4. Explain how the following can be detected: (a) leaking fuel valve, (b) leaking exhaust valve, (c) leaking air inlet valve.

5. What could be the possible causes of failure of cooling water supply in an auxiliary diesel engine? In the event of such a failure and piston seizure, explain how you would put the auxiliary back into service.

6. Describe how a replacement crankshaft timing chain is fitted to a diesel engine. Explain how the correct tension is achieved and the camshaft timing checked.

7. Sketch and describe a fuel supply system for a diesel engine showing all heaters, filters, pumps, etc., describe how oil temperature is controlled.

8. Explain how an explosion can occur in an air start line and state the precautions to be taken to prevent a re-occurrence.

9. Explain how the following are detected: (a) choked fuel valve, (b) early injection of fuel, (c) leaking piston rings, (d) after burning.

10. Describe with the aid of sketches how a large marine diesel engine is reversed.

11. Enumerate the causes of piston overheating and failure and the way in which these faults are detected. Describe how you would examine a piston after removal.

12. Discuss the indications of: (a) a scavenge fire, (b) an overheated piston. What would be your actions subsequent to these discoveries?

13. Detail examination and tests to be carried out before the commencement of a voyage on a diesel engined vessel. Discuss the routine adopted on your last vessel.

14. Describe how you would prepare a main bearing for examination. What defects would you look for and how would you remove the bottom half of the bearing?

15. Describe the overhauling of a two-stage air compressor. How are the correct running clearances ensured on bearings and cylinders? State the tests and precautions necessary before the compressor is put back into service.

16. Enumerate the fittings, and explain their purpose, on the shell of a main starting air receiver. Explain how they are attached. State the precautions to be observed when filling, emptying, opening up, cleaning and applying protective compounds to the internal surfaces.

17. Describe, with the aid of a line sketch, the lubricating oil system of supply to the main diesel engine of your last vessel. Discuss the routine tests and operational procedures adopted to ensure that the oil reached the engine in optimum condition.

18. Discuss the fuel injection timing of a diesel engine at full and light load conditions. Explain how a reduction in the amount of fuel injected is achieved with: (a) mechanically operated fuel valves, (b) hydraulically operated fuel valves.

19. Describe a method of determining the clearance of a main engine connecting rod top end bearing. Why are the clearances for top and bottom end bearings generally different?

20. Explain how individual cylinder powers in a medium speed auxiliary diesel engine can be balanced. Comment on the effects of running the engine for extended periods in an unbalanced condition.

21. Discuss the reasons for an engine failing to pick up speed and run when fuel is put on, after the engine has been turned on air.

22. Describe the inspections that you would carry out and the reasons for same during a one day stay in port after a long sea voyage, confine your discussion to the main engine only.

23. What could be the possible causes of a main diesel engine slowing down from full speed without any alteration in the control lever settings? Explain your course of action in each case.

24. Sketch and describe how a main engine piston is cooled, discuss the advantages and disadvantages of the cooling medium you have chosen. What are the possible causes of high temperature piston cooling returns?

25. What are the desirable conditions for the combustion of fuel in the cylinder of a main diesel engine. Give a typical analysis of the exhaust gas from the engine.

First Class

1. Liner wear has been said to be proportional to sulphur content of fuel and inversely proportional to liner temperature. Discuss this statement and also the effects of running for prolonged periods on low load.

2. Cylinder heads, liners and pistons have failed due to thermal stresses. Discuss how this problem is overcome in the design stages of the engine and when operating the engine.

3. Sketch and describe the main engine crosshead bearing and guide for a large slow running diesel showing how the connecting and piston rods are attached. Discuss the bearing clearances and explain the significance of 'mirror finish' and how adequate lubricating oil pressure for the top end bearing is achieved.

4. Instances have occurred of crankcase explosions, with serious results in some cases. What do you consider may have been the cause, or causes, of these explosions? What measures would you suggest with a view to minimising the danger of explosion? Describe a device which would give warning of dangerous conditions in the crankcase.

5. Gas turbines have recently been used as main propulsion units in a number of ships. Sketch and describe such an arrangement and comment on the advantages and disadvantages of this form of propulsion system.

6. Explain why it is advantageous to control the cooling media in a main diesel engine within close temperature limits. Sketch and describe a system for controlling the temperature automatically.

7. Describe how you would organise and make a detailed examination of the bedplate and crankshaft of a main engine. What defects might you expect to find and what could be the causes of the defects?

8. Discuss the importance of timing cylinder lubrication and where the oil should ideally be injected. Sketch and describe a cylinder quill for the injection of oil.

9. With reference to a main engine discuss the following: (a) maintaining a low level of lubricating oil in the engine sump, (b) running Ahead and braking with Astern air, (c) carrying on running the engine with a scavenge fire.

10. Where alloyed aluminium pistons are used in medium speed diesels discuss the following: (a) incendive sparking, (b) position of oil scraper rings, (c) piston and liner clearances, (d) piston lubrication and cooling.

11. Sketch and describe a method by which the auxiliary pumps necessary to the running of a medium speed diesel are operated by the engine. Discuss the advantages and disadvantages of this arrangement and provisions made for stand-by periods.

12. Discuss the factors which govern mean piston speed in a diesel engine.

13. Sketch and describe a system of governing medium speed diesel engines which are geared together and drive a controllable pitch propeller. Explain how governing takes place.

14. Discuss the following items and where appropriate enumerate the advantages and disadvantages: (a) chrome plating of cylinder liners, (b) plating of piston rings (c) cylinder liner wear, (d) piston ring breakage.

15. Some of the items in the crankcase of a marine diesel engine are boundary lubricated. Discuss a typical example and the measures taken to minimise the effects of boundary lubrication for the example of your choice.

16. You are appointed Chief Engineer of the first of a series of motorships fitted with large bore diesel engines. State what possible troubles may be encountered, enumerate defects that may manifest themselves during normal maintenance and discuss the reasons for these possible troubles and defects.

17. Give possible causes of explosions in crankcases and air start lines. Discuss also the probable causes of fires in the exhaust manifold of an auxiliary diesel and in the scavenge spaces of the main engine.

18. Discuss critically the effects on cylinder maximum pressure, compression pressure, exhaust temperature, power and engine speed due to a change in: (a) fuel temperature, (b) relative density of fuel, (c) turbo-charge temperature and pressure, (d) ambient temperature.

19. Explain how the following occur in a marine diesel engine when burning residual fuel and explain how they can be prevented or alleviated: (a) fuel valve nozzle deposits, (b) sticking piston rings, (c) cylinder liner corrosion, (d) crankcase lubricating oil contamination.

20. Discuss the various types of stresses that may arise in the crankshafts of large marine diesel engines and explain what aspects of design and maintenance assist in alleviating them.

21. Discuss the manner in which fuel oils are injected into the cylinders of large marine diesel engines. What factors affect the

efficient burning of the fuel? Draw a diagram for an engine of your choice showing the period of fuel injection at full power and explain the effects of advancing the timing on: (a) specific fuel consumption, (b) cylinder pressures, (c) cylinder temperatures and exhaust temperatures.

22. Sketch and describe an automatic boiler which may be used for heating and domestic purposes. Enumerate all the important mountings and discuss the manner in which the boiler is automatically controlled.

23. Sketch and describe an air starting valve for a large marine diesel engine. Draw the timing diagram for the engine whose starting valve you have chosen and discuss the effects of the valve sticking open.

24. Explain, with respect to a mechanical governor as fitted to an auxiliary diesel generator, how the following terms may be applied: (a) error signal, (b) feed back, (c) sensing element, (d) closed loop.

25. What is meant by viscosity of fuel oil and how is it affected by temperature? Discuss the modifications and additions necessary to the fuel injection system of a large marine diesel engine in order that it may burn, efficiently, fuel of high viscosity and lower calorific value.

INDEX

REED'S MARINE ENGINEERING SERIES

The series covers the full range of subjects for all grades of the Department of Trade Certificates of Competency in Marine Engineering. The books will be extremely useful to marine engineer cadets and other engineering students studying on Technical Education Council and SCOTEC Courses. Material is presented from first principles with many diagrammatic sketches and worked solutions to examples.

Vol. 1. MATHEMATICS.
Vol. 2. APPLIED MECHANICS.
Vol. 3. HEAT & HEAT ENGINES.
Vol. 4. NAVAL ARCHITECTURE.
Vol. 5. SHIP CONSTRUCTION.
Vol. 6. BASIC ELECTROTECHNOLOGY.
Vol. 7. ADVANCED ELECTROTECHNOLOGY.
Vol. 8. GENERAL ENGINEERING KNOWLEDGE.
Vol. 9. STEAM ENGINEERING KNOWLEDGE.
Vol. 10. INSTRUMENTATION & CONTROL SYSTEMS.
Vol. 11. ENGINEERING DRAWING.
Vol. 12. MOTOR ENGINEERING KNOWLEDGE.

Published by:

THOMAS REED PUBLICATIONS LIMITED

36-37 Cock Lane, London EC1A 9BY